Ethnobotany and Uses of Native Plants in the Bosque by American Indian Tribes of the Southwest: An Integrated/Transdisciplinary Approach to Ecosystem Services

Written By: Tyler Pounds (formerly Tyler Walborn)

Illustrated By: Tyler Pounds (formerly Tyler Walborn)

Edited By: Todd Walborn

Front and Back Cover Artwork (Photographs) Taken By: Kimberly Roy

Front and Back Cover Artwork By: Tyler Pounds (formerly Tyler Walborn)

Table of Contents

Copyright Notice

Copyright © 2023 Tyler Pounds

To request permissions, contact Mr. Pounds at

Email: tylerpounds1989@icloud.com

Follow me on these social media channels!

Facebook:
https://www.facebook.com/profile.php?id=100088237461903

Linkedin:
https://www.linkedin.com/in/tyler-pounds-87798125b/

Twitter:
https://twitter.com/tylerpounds1989

Instagram:
https://www.instagram.com/tpounds04/

Youtube:
https://www.youtube.com/channel/UCpQIY6TbJ0csL8r-lqH7PQQ

Goodreads.com:
https://www.goodreads.com/author/show/47839486.Tyler_Pounds

PAPERBACK BOOK ISBN: 979-8-8692-3810-8 (IngramSpark)

Library of Congress Control Number: 2023911516

This book's *City of Publication* is in Medford, Oregon. This book was written as a self-published work using "IngramSpark".

Dedication(s)

IN HONOR OF E.O. WILSON

Thank you for giving me the encouragement needed to continue my research after I thought all hope was lost.

Greetings from a fellow Alabama native and a Ecological Anthropologist/Ethnobiologist

Label: DeletedItems (30 days) Expires: Sun 5/3/2020 7:55 PM

Wilson, Edward O. <ewilson@oeb.harvard.edu>
Fri 3/6/2020 3:45 PM
Tyler Pounds; Wilson, Edward O. <ewilson@oeb.harvard.edu>

Dear Mr. Pounds,

Thank you very much for your letter of February 21 and the copy of your interesting and important articles. I've read this work, which is excellent, and I wish you all success as you proceed in thought and research.

With warm regards,

Edward O. Wilson

EOW:kmh

Professor Edward O. Wilson | Harvard University | Museum of Comparative Zoology | 26 Oxford Street | Cambridge MA 02138-2902

RIP
E. O. Wilson (1929-2021)

IN HONOR OF CHARLES DARWIN

Charles Darwin was a Biological Anthropologist who studied human origins and applied his theories of "Evolution" to wildlife species and to humans. Later on, many of Darwin's theories would be applicable and applied to both the fields of Anthropology and Biology. Darwin would also occasionally be employed as a "Geologist" from time to time. I honor his work (which seemed to be just as integrated and transdisciplinary as mine).

RIP
Charles Darwin (1809-1882)

Forward and Acknowledgements

I am writing this book today with encouragement from the Native American Tribes of the Southwest, and for the honoring of Indigenous Peoples all over the world, especially for the Indigenous Peoples of the American Southwest. All traditional medicines, knowledge bases, and customs associated with the fields of Ethnobotany, Ethnobiology, and Ecological Anthropology should be respected and preserved for the safekeeping of American Indian culture for generations to come. *Indigenous people and their traditional ecological knowledge base systems (TEK) are helping to preserve over 80 percent of the world's biodiversity. Maryellen Kennedy Duckett, National Geographic, "Nature Needs us to Act-Now," March 4th, 2020.* Knowing this fact, hopefully more Western scientists will listen to Indigenous Peoples, as they are the key to helping the human race, and the world, avoid extinction.

I would like to honor all of the ancestors that came before us. They continually gift us all with the knowledge of the plants and medicines of the earth. I would also like to thank my ancestors, friends, and Indigenous People for their continued support in the process of me writing this book. You all gave me the strength, intellect, and passion to carry on with this endeavor. Together we can save the earth.

Traditional Ecological Knowledge

In this book, we will discuss the ethnobotanical uses of plants and how people (humans) use the plants in the Bosque bio-region of the Southwestern United States (American Southwest). Humans have evolved over thousands of years in different ecosystems and environments, during which time they mastered their surroundings. This phenomena or knowledge of ecosystems from living and directly interacting with various ecosystems in a particular region over hundreds to thousands of years is known as "traditional ecological knowledge," or TEK. In this particular case, we will learn how different American Indian tribes utilized certain species of plants that still grow naturally and commonly in and around the Bosque bio-region. Traditional Ecological Knowledge (TEK) is useful for people to know, because it shows us which plants and animals are important to various cultures all over the world. Sometimes it is for medicine, sometimes it is for food or clothing, but various other uses also exist.

In this way, we gain an important, but different, insight into how and why certain plants and animals are important to humans in various parts of the world. Knowing and acknowledging TEK systems can also certainly increase conservation efforts for certain species of plants and animals. It gives people reasons of why these species are important to us, and therefore we are more likely to save the ecosystems which we so heavily depend upon. It should also be noted that TEK is "localized knowledge" within a local community or region. Therefore, TEK will be limited to local plant and animal species (when working with Indigenous communities it is important to remember this). I am listing a link below for Traditional Ecological Knowledge, and how it is being applied by real-world scientists. It is a must read!

U.S. Fish & Wildlife Service, "Traditional Ecological Knowledge for Application by Service Scientists," Ninepipe National Wildlife Refuge of Montana, October 2, 2018
https://www.fws.gov/nativeamerican/pdf/tek-fact-sheet.pdf

The term Traditional Ecological Knowledge (TEK) is any system of knowledge that is related to plant, animal, or ecosystem use. This knowledge is generally acquired through applied knowledge and practices that any group of people have utilized for any time period between a few hundred to a few thousand years between themselves and the natural world around them. Generally, the interactions between humans and wildlife resources is what drives this acquired knowledge. Interactions between wildlife resources and humans generally include (but are not limited to) hunting, fishing, trapping, agriculture, spiritual ceremonies and usages, traditional clothing and regalia making, etc. TEK is also directly intertwined with human culture, which largely drives how people interact with nature, and thus the fields of "Ecological Anthropology," "Ecosystem Services," and "Ethnobiology" are perfect scientific approaches to study such systems of plants, animals, and ecosystems.

U.S. Department of the Interior. (2020, August 5). *Overview of TEK*. National Parks Service. Retrieved October 3, 2022, from https://www.nps.gov/subjects/tek/description.htm

Hoagland, S. J. (2016). Integrating Traditional Ecological Knowledge with Western Science for Optimal Natural Resource Management. *IK: Other Ways of Knowing, 3*(1), 1–15. https://doi.org/https://doi.org/10.18113/P8ik359744

Berkes, F. 1999. *Sacred Ecology: Traditional Ecological Knowledge and Resource Management*. Philadelphia, PA: Taylor and Francis Publishing Company.

Applications of TEK through Methodological Practices

Indigenous Ways of Knowing (IWOK) is applied "kinesthetic learning," or "learning by doing." Traditional Ecological Knowledge (TEK) is accumulated over hundreds, if not thousands, of years through figuring out what works best by mastering one's environment. Think about agricultural practices. It took humans thousands of years to figure out what worked best in order to grow our crops as quickly and efficiently as possible. We can also imagine how long it took humans to figure out what combinations of crops grew together efficiently without degrading the soils in which they grew. This is a prime example of how both Indigenous Ways of Knowing (IWOK) and Traditional Ecological Knowledge (TEK) are applied through "kinesthetic learning," or "learning by doing," in order to address real-world problems. This also shows how both

Traditional Ecological Knowledge (TEK) and Indigenous Ways of Knowing (IWOK) are utilized in order to further human society.

Participatory Action-Based Research Approach

"Participatory Action Research (PAR) is an approach to inquiry which has been used since the 1940s. It involves researchers and participants working together to understand a problematic situation and change it for the better. There are many definitions of the approach, which share some common elements. PAR focuses on social change that promotes democracy and challenges inequality; is context-specific, often targeted on the needs of a particular group; is an iterative cycle of research, action, and reflection; and often seeks to 'liberate' participants to have a greater awareness of their situation in order to take action. PAR uses a range of different methods, both qualitative and quantitative."

Macbeth, S. (n.d.). Participatory Action Research. Participatory Action Research | Participatory Methods. Institute of Development Studies. Retrieved September 25, 2022, from *https://www.participatorymethods.org/glossary/participatory-action-research*

This form of "inquiry" has mainly been used in conservation education, as well as in the social sciences, in order to address a wide range of social and ecological issues that face large communities/tribes of people. Zoos and aquariums often utilize this approach in conservation education in order to actively engage their communities, as a whole, in order to promote critical thinking skills for issues that their respective communities face. This is for researchers to figure out how the issues could be effectively addressed, while still being actively engaged and involved with local communities. This form of scientific inquiry integrates science with local populations of people so they do not feel left out. This form of scientific inquiry also tends to "recruit," so to speak, for the fields of science within STEM. It gets people (especially younger generations) excited about science, and helps to bridge the gaps between local communities and science projects, as a whole.

Researchers will need to meet with local community and tribal members in order to first identify the ecological and social issues at hand. Generally, researchers need to be quiet and listen to the local community and tribal members in order to properly identify said issues. Then they can figure out how to best approach and solve these current issues. Being actively engaged with community members has been largely successful and makes the community members more likely to trust the scientific inquiry that is being conducted. I personally feel that more research needs to include this form of "action-based inquiry."

Community-Based Service Learning

Community-based service learning is another invaluable tool that can be used in conjunction with "participatory action-based research methodologies." This is a tool that combines classroom instruction with real-world application through a community service-based research project, which allows both the students and the local communities and tribes to see firsthand how the classroom instruction and subject matter can be applied in the real world. Community-based service learning also bridges the gap between local scientists and community and tribal leaders. This will help to build and strengthen those relationships in order to further the field of science, as well as promoting the needs of the communities and tribes that need certain issues addressed.

It should also be noted that "service scientists" practicing community-based service learning should have continued interactions with the various communities in which they serve. Quite commonly, most Western scientists come into a community, conduct their scientific research, and then leave (never to return or be seen by the community again). This causes distrust between Indigenous communities and the scientific community, as a whole. Indigenous communities like to see continued involvement long after scientific projects have been completed. That way, these communities feel that they are being taken seriously, and not being taken advantage of by the Western world. When it comes to community-based service learning projects, the best types of relationships are "continued relationships" (especially when dealing with Indigenous communities). Continued involvement should be the standard for community-based service learning.

This form of scientific inquiry should help to promote "decolonizing biodiversity," and how science should be conducted when interacting with Indigenous People. We should all remember that for most Indigenous People, trust is a huge aspect of community structure. We should also remember that

Indigenous communities all over the world have been practicing their Traditional Ecological Knowledge (TEK) bases for thousands of years. However, these knowledge bases are almost exclusively localized and relevant to the communities in which they live. It is important for scientists to understand that TEK is a "community-based/-oriented science." This being said, having positive, continued relationships with Indigenous communities in which you plan on serving is a must.

"Community-based learning refers to a wide variety of instructional methods and programs that educators use to connect what is being taught in schools to their surrounding communities, including local institutions, history, literature, cultural heritage, and natural environments. Community-based learning is also motivated by the belief that all communities have intrinsic educational assets and resources that educators can use to enhance learning experiences for students."

The Glossary of Education Reform. (2014, March 3). Community-Based Learning. Community-Based Learning Definition. Retrieved September 25, 2022, from https://www.edglossary.org/community-based-learning/

"Field-based "experiential learning" with community partners is an instructional strategy—and often a required part of the course. The idea is to give students direct experience with issues they are studying in the curriculum and with ongoing efforts to analyze and solve problems in the community. A key element in these programs is the opportunity students have to both apply what they are learning in real-world settings and reflect in a classroom setting on their service experiences. These programs model the idea that giving something back to the community is an important college outcome, and that working with community partners is good preparation for citizenship, work, and life."

Kuh, G. D. (2008). High-Impact Educational Practices: What They Are, Who Has Access to Them, and Why They Matter, Service Learning/Community-Based Learning, Association of American Colleges and Universities. Retrieved October 21, 2022, from https://secure.aacu.org/AACU/PubExcerpts/HIGHIMP.html#:~:text=In%20these%20programs%2C%20field%2Dbased,solve%20problems%20in%20the%20community

The Importance of Plants in Tangible and Intangible Cultural Heritage, Heritage Resources, and Natural Resource Management

"'Tangible Cultural Heritage' refers to physical artefacts produced, maintained, and transmitted intergenerationally in a society. It includes artistic creations, built heritage such as buildings and monuments, and other physical or tangible products of human creativity that are invested with cultural significance in a society. 'Intangible Cultural Heritage' indicates 'the practices, representations, expressions, knowledge, skills – as well as the instruments, objects, artefacts, and cultural spaces associated therewith – that communities, groups, and, in some cases, individuals recognize as part of their Cultural Heritage" (UNESCO, 2003). Examples of intangible heritage are oral traditions, performing arts, local knowledge, and traditional skills.

Tangible and intangible heritage require different approaches for preservation and safeguarding, which has been one of the main motivations driving the conception and ratification of the 2003 UNESCO Convention for the Safeguarding of the Intangible Cultural Heritage. The Convention stipulates the interdependence between Intangible Cultural Heritage, Tangible Cultural Heritage, and Natural Heritage, and acknowledges the role of Intangible Cultural Heritage as a source of cultural diversity and a driver of sustainable development. Recognizing the value of people for the expression and transmission of Intangible Cultural Heritage, UNESCO spearheaded the recognition and promotion of living human treasures, "persons who possess to a very high degree the knowledge and skills required for performing or recreating specific elements of the Intangible Cultural Heritage."

Tangible and Intangible Cultural Heritage, Riches Resources, Published: November 27th, 2014. Accessed: August 9th, 2021
https://resources.riches-project.eu/glossary/tangible-and-intangible-cultural-heritage/

Additional Sources:

UNESCO (2003) Convention for the Safeguarding of the Intangible Cultural Heritage. Paris: UNESCO.

UNESCO (n.d.) Guidelines for the Establishment of National "Living Human Treasures" Systems. Paris: UNESCO.

Plants are crucially important when talking about Tangible and Intangible Cultural Heritage, especially when referring to Indigenous cultures all over the world. Let's use Native American culture as an example. Think about "Tangible Cultural Heritage," and how Native Americans have used plants and various plant fibers to make baskets, sew clothing with, create corn husk dolls, construct teepees with various tree wood poles, construct sweat lodges with willows, etc. Don't you think it would be important to preserve not only the natural resources of the plant, itself (especially if the plant is known to be an endangered species), but to also document the exact uses of the plant in order to conserve human culture? It is crucial that scientists from various backgrounds understand the implications and values of having such knowledge from the field of Ethnobotany. In other words, natural resource managers will have proof of human consumption of certain plant species by the documenting of plant harvesting, traditional plant gathering, and plant utilization through cultural usages. Also, anthropologists and archaeologists will also have proof of tangible heritage through the cultural usages of certain plant species and therefore will be able to gain rights and have access to certain plant species for culturally related purposes.

Now let's think about "Intangible Cultural Heritage," and how Native Americans have used plants in order to create objects and artefacts associated with traditional ceremony. These uses include making pipes to smoke with, the utilization of plants for smudging (a ceremonial practice believed to cleanse spiritual energies), preparing kinnikinnick (a specialized smoking blend of certain plants for use in pipe ceremonies), the making of certain spiritual objects, such as bullroars and wands (Apache Tribes), etc. Documenting the uses of plants is also critical for the preservation of cultural heritage, so tribal members and tribal descendants will always have access to knowledge of ceremonies and spiritual practices (in case this knowledge is ever lost). I wholeheartedly encourage tribes to create an archival list of plants and plant species that are known to be used for both tangible and intangible heritage uses. I also believe the tribes "fish & game departments" need to compile their own data and archives of such species, so as to prioritize them when it comes to natural resource conservation for cultural usages. This will not only save Native American cultures, but will also give the United States Government proof of how and why it is important to preserve and conserve such plants and botanical species for future cultural uses. I have created my own forms for the gathering and dissemination of ethnobotanical data to be used by social scientists and natural resource managers. These forms, which can be found below, are for the collection of Traditional Ecological Knowledge (TEK), in regards to the preservation of cultural heritage (both tangible and intangible), but also for the preservation of natural resources and the field of wildlife conservation.

TEK Forms, the BIOTEK Database Information Management System, the TRIBAL-RAD TEK System, and the GOV-RAD TEK System

<u>Traditional Ecological Knowledge (TEK) Botanical Data Gathering Sheet</u>

For use by cultural resource managers such as Anthropologist's and Archaeologist's

Tribe Or Community Being Studied:

Plant Species Common Name:

Plant Species Latin (Scientific Name):

<u>Plant Species Uses</u>

Cultural Uses:

Spiritual Uses:

Medicinal Uses:

Of Different Cultural Uses For Said Plant Species Of Interest:

Of Different Spiritual Uses For Said Plant Species Of Interest:

Of Different Medicinal Uses For Said Plant Species Of Interest:

Plant Species Utilized Under "Tangible Heritage":

Plant Species Utilized Under "Intangible Heritage":

Plant Species Utilized As "Naturopathic Medicine":

<u>Traditional Ecological Knowledge (TEK) Botanical Data Gathering Sheet</u>

For use by natural resource managers in the various fields of conservation

Tribe Or Community Being Studied:

Plant Species Common Name:

Plant Species Latin (Scientific Name):

Plant Species Designated "Plant Family":

Plant Species Habitat Area Or Range:

Conservation Status (Please Circle One): EX-Extinct, EW-Extinct in Wild, CR-Critically Endangered, EN-Endangered, VU-Vulnerable, NT-Near Threatened, LC-Least Concern

<u>Species Richness Level</u>

Alpha Diversity (richness and evenness of individuals within a habitat unit):

Beta Diversity (expression of diversity between habitats):

Gamma Diversity (landscape diversity or diversity of habitats within a landscape or region):

Recommendations For Plant Species Of Interest:

There is the potential for both the various Native American tribes and governmental agencies to establish various databases in regards to Traditional Ecological Knowledge (TEK) and Cultural Resource Management (CRM), using my above forms to keep track of the various biological and cultural usages that plants have within our various societies (both tribal and non-tribal, alike). I am going to call this theoretical database that I would like to see created for both the tribes and various governmental agencies to use "BIOTEK," which stands for Biological Traditional Ecological Knowledge Database. I am going to preview a template of how the databases should be formatted for proper storage of all ethnobotanical data concerning how plants are managed and utilized in the ecosystems (Natural Resource Management and Ecosystem Services), but also how plants relate to culture and how various cultures utilize the plants for cultural preservation and heritage purposes (Cultural Resource Management).

BIOTEK Database For Cultural Resource Management

Tribe Or Community Being Studied:	Plant Species Common Name:	Plant Species Latin (Scientific Name):	Plant Species Cultural Uses:	Plant Species Spiritual Uses:	Plant Species Medicinal Uses:	# Of Different Cultural Uses For Said Plant Species Of Interest:	# Of Different Spiritual Uses For Said Plant Species Of Interest:	# Of Different Medicinal Uses For Said Plant Species Of Interest:	Plant Species Utilized Under "Tangible Heritage":	Plant Species Utilized Under "Intangible Heritage":	Plant Species Utilized As "Naturopathic Medicine":

BIOTEK Database For Natural Resource Managers

Tribe Or Community Being Studied:	Plant Species Common Name:	Plant Species Latin (Scientific Name):	Plant Species Designated "Plant Family":	Plant Species Habitat Area Or Range:	Conservation Status: EX-Extinct, EW-Extinct in Wild, CR-Critically Endangered, EN-Endangered, VU-Vulnerable, NT-Near Threatened, LC-Least Concern	Species Richness Level: Alpha Diversity (richness and evenness of individuals within a habitat unit)	Species Richness Level: Beta Diversity (expression of diversity between habitats)	Species Richness Level: Gamma Diversity (landscape diversity or diversity of habitats within a landscape or region)	Reccomen. For Plant Species Of Interest

It should be noted that such databases could house information for critically endangered species or cultural practices that should, therefore, be protected. I recommend that once all tribal and governmental BIOTEK databases are established, they be protected with encrypted log-ins, as well as government or tribal IDs, most of which have microchips that a person can place into a microchip encryption data reader. This is both safe and practical in order to protect and house pertinent and important data relating to both governmental and tribal information. It is crucial to the continuation of scientific discovery for the continuation of all Traditional Ecological Knowledge (TEK) to be housed and documented in safe and ethical manners. I also believe that both the tribes and the governmental agencies that manage natural resources and cultural heritage should integrate the BIOTEK databases with modern Geographic Information Systems (GIS) maps and mapping applications. I believe that with the forms I have created above in order to collect data, alongside the BIOTEK database templates I have listed above, that both tribal and governmental agencies will be able to integrate GIS maps and mapping technology to fully save, manage, and conserve both cultural heritage and natural resources concerning ethnobotanical TEK data. Now, before diving into the potential GIS mapping applications that could be integrated with these BIOTEK databases, let's discuss the TRIBAL-RAD TEK Systems & GOV-RAD TEK Systems.

The TRIBAL-RAD TEK System:

This stands for "Tribal-Ready, Available, Data concerning Traditional Ecological Knowledge System." I propose that the TRIBAL-RAD TEK System be a larger computer server/database system that can house my BIOTEK (Biological Traditional Ecological Knowledge) database System. The much larger TRIBAL-RAD TEK System server can also house specific tribal fish and wildlife laws, codes, emergency contacts, fish/wildlife contacts, any news or information that relates to Traditional Ecological Knowledge (TEK), or any other knowledge that promotes cultural heritage through the fields of Cultural Resource Management (CRM) and Natural Resource Management.

The GOV-RAD TEK System:

This stands for "Government-Ready, Available, Data concerning Traditional Ecological Knowledge System." I propose that the GOV-RAD TEK SYSTEM be a larger computer server/database system that can house my BIOTEK (Biological Traditional Ecological Knowledge) database System. The much larger GOV-RAD TEK System server can also house specific government fish and wildlife laws, codes, emergency contacts, natural resource agency contacts (such as BLM, Forestry Service, Fish/Wildlife Service, National Park Service, etc.), any news or information that relates to Traditional Ecological Knowledge (TEK), or any other knowledge that promotes cultural heritage through the fields of Cultural Resource Management (CRM) and Natural Resource Management. Now, let's discuss how GIS mapping technologies can be integrated with the BIOTEK Database Information Management System, the TRIBAL-RAD TEK System, and the GOV-RAD TEK System. I am going to give you all an example of how GIS Mapping technology could be integrated within all three systems. Let's look at my tribal mapping research project for the state of Oklahoma and think about how this type of data could be used within TEK Databases (both tribal and governmental).

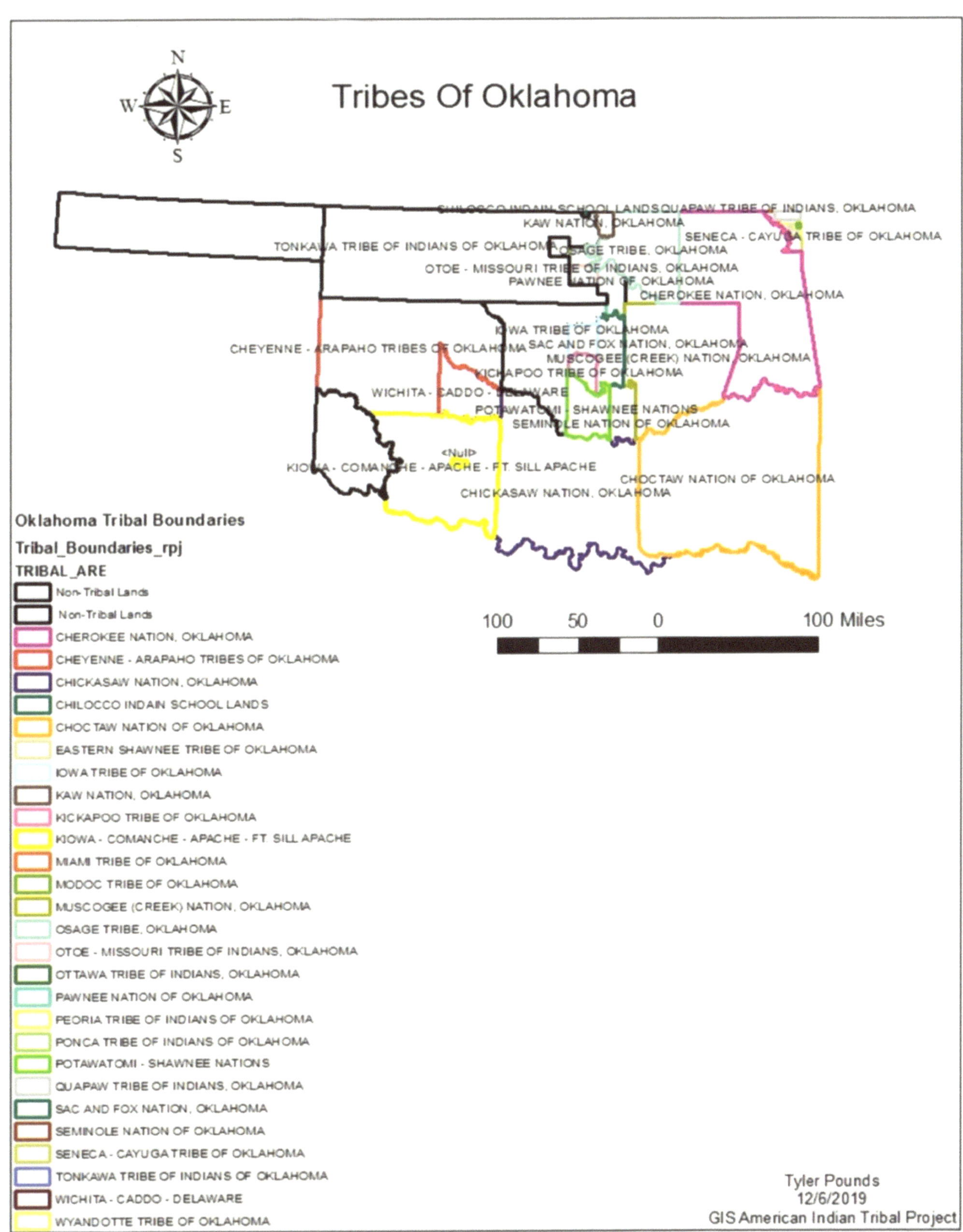

Tyler Pounds-New Mexico State University-MA Anthropology (Ecological)

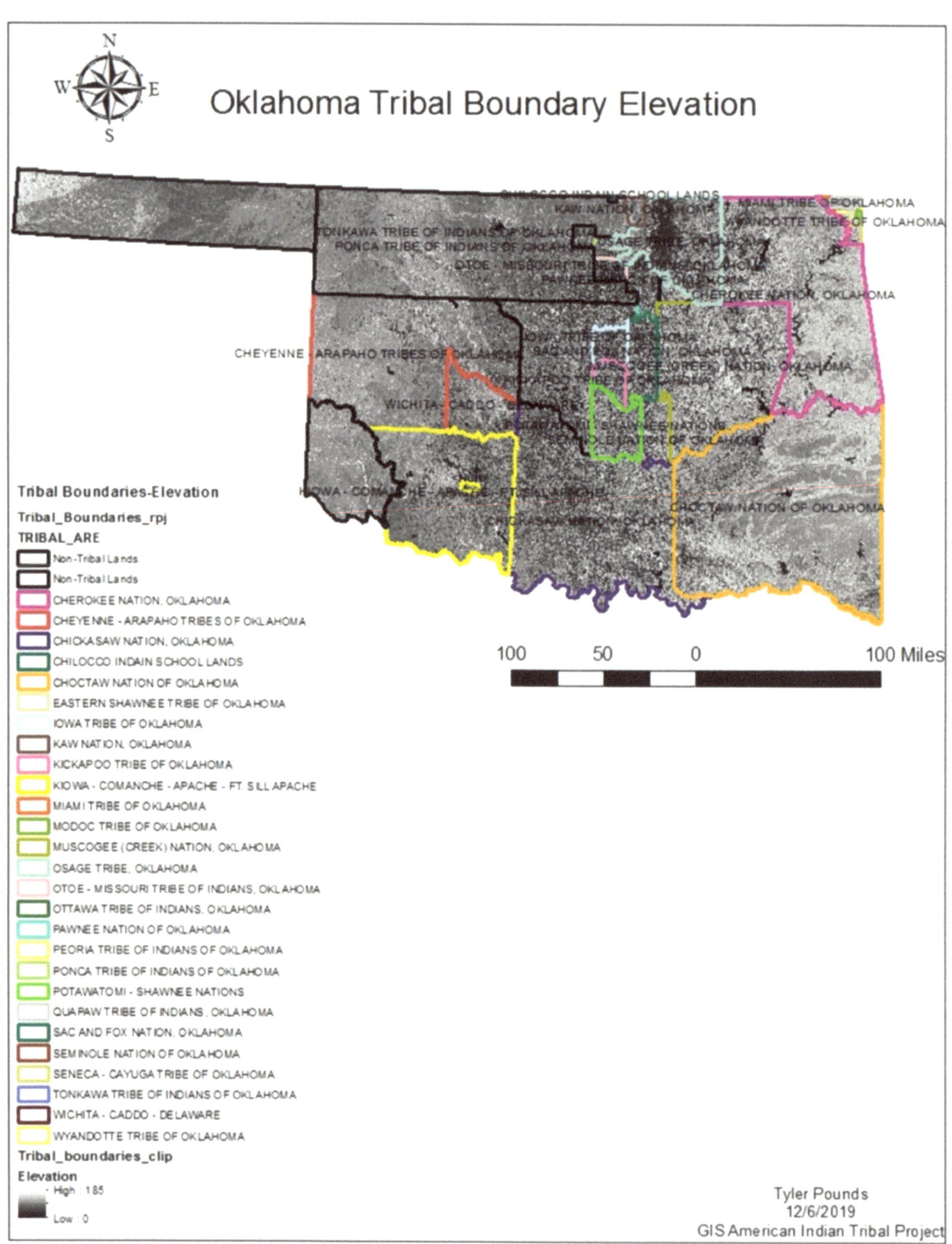

Tyler Pounds-New Mexico State University-MA Anthropology (Ecological)

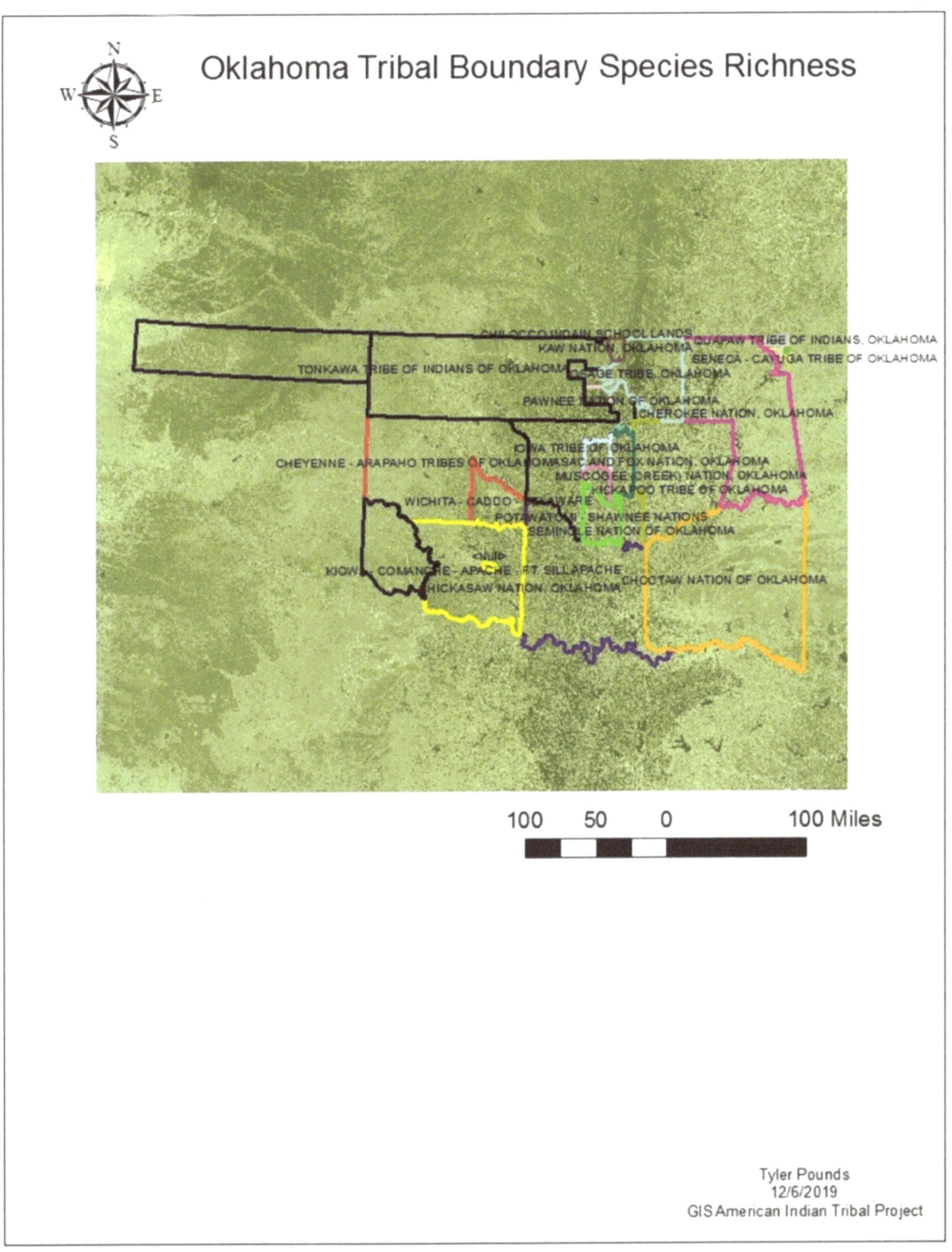

Tyler Pounds-New Mexico State University-MA Anthropology (Ecological)

Let us now look at the data and the results of what has been analyzed. Before we begin, it should be noted that we used "Majority" data in order to determine the number of species (species richness) in various tribal boundary areas. This is the original data-set in which we had to analyze the results for all of the tribes in Oklahoma. The original data-set can be found below.

OBJECTID	TRIBAL_ARE	COUNT_	AREA	MIN_	MAX_	RANGE	MEAN	STD	SUM_	VARIETY	MAJORITY	MINORITY	MEDIAN
1	OTTAWA TRIBE OF INDIANS, OKLAHOMA	69176	62258400	24	168	144	116.8039	17.17582	8080025	124	117	54	118
2	WYANDOTTE TRIBE OF OKLAHOMA	95569	86012100	26	169	143	117.1499	17.60695	11195895	126	120	67	119
3	SEMINOLE NATION OF OKLAHOMA	1652079	1.49E+09	25	173	148	102.3801	30.82949	1.69E+08	146	65	30	116
4	WICHITA - CADDO - DELAWARE	3320218	2.99E+09	22	166	144	97.88982	23.20741	3.25E+08	142	98	22	99
5	CHICKASAW NATION, OKLAHOMA	21229785	1.91E+10	20	182	162	93.05466	29.55928	1.98E+09	163	58	21	101
6	CHILOCCO INDAIN SCHOOL LANDS	38205	34384500	32	165	133	96.55307	20.26152	3688810	112	85	35	90
7	CHEYENNE - ARAPAHO TRIBES OF OKLAHOMA	19467243	1.75E+10	0	167	167	103.9543	18.23664	2.02E+09	150	113	167	106
8	KIOWA - COMANCHE - APACHE - FT. SILL APACHE	13379511	1.2E+10	0	176	176	95.27787	23.15684	1.27E+09	159	109	176	100
9	PAWNEE NATION OF OKLAHOMA	1284112	1.16E+09	25	170	145	86.43124	30.60335	1.11E+08	145	58	170	67
10	PEORIA TRIBE OF INDIANS OF OKLAHOMA	108598	97738200	24	159	135	109.9457	15.32308	11939881	117	117	41	112
11	MIAMI TRIBE OF OKLAHOMA	116582	1.05E+08	24	167	143	107.953	19.72529	12585377	129	112	28	112
12	SENECA - CAYUGA TRIBE OF OKLAHOMA	234141	2.11E+08	22	159	137	108.6517	28.67657	25439827	128	123	56	117
13	TONKAWA TRIBE OF INDIANS OF OKLAHOMA	412909	3.72E+08	25	167	142	96.33118	18.4576	39776013	137	95	30	94
14	PONCA TRIBE OF INDIANS OF OKLAHOMA	470651	4.24E+08	24	176	152	94.37269	26.36619	44416603	151	93	28	95
15	EASTERN SHAWNEE TRIBE OF OKLAHOMA	61452	55306800	26	156	130	114.7326	14.03094	7050546	107	117	33	117
16	OTOE - MISSOURI TRIBE OF INDIANS, OKLAHOMA	580817	5.23E+08	22	171	149	75.74002	27.13076	43991091	147	58	32	61
17	MUSCOGEE (CREEK) NATION, OKLAHOMA	13989128	1.26E+10	16	180	164	105.9043	30.19446	1.48E+09	162	121	180	117
18	SAC AND FOX NATION, OKLAHOMA	2161664	1.95E+09	24	165	141	92.31711	32.10907	2E+08	139	57	32	97
19	IOWA TRIBE OF OKLAHOMA	1035944	9.32E+08	25	163	138	89.31103	30.26286	92521221	135	57	27	94
20	KICKAPOO TRIBE OF OKLAHOMA	922158	8.3E+08	25	166	141	92.56739	30.28596	85361760	140	63	32	101
21	POTAWATOMI - SHAWNEE NATIONS	2571018	2.31E+09	23	176	153	98.31924	30.68375	2.53E+08	150	60	176	113
22	CHOCTAW NATION OF OKLAHOMA	31395705	2.83E+10	16	183	167	109.2094	21.99849	3.43E+09	168	114	183	113
23	QUAPAW TRIBE OF INDIANS, OKLAHOMA	253442	2.28E+08	24	160	136	105.1928	20.55641	26660272	130	117	150	110
24	CHEROKEE NATION, OKLAHOMA	20015792	1.8E+10	16	184	168	111.156	25.8969	2.22E+09	166	123	183	118
25	OSAGE TRIBE, OKLAHOMA	6630598	5.97E+09	23	175	152	88.58196	29.5686	5.87E+08	152	58	175	86
26	KAW NATION, OKLAHOMA	448034	4.03E+08	23	148	125	95.53806	22.39327	42804300	124	84	31	98
27	MODOC TRIBE OF OKLAHOMA	21089	18980100	48	144	96	115.6614	10.2785	2439183	74	111	68	116

From the data-set that is listed above, we were able to analyze what tribes in Oklahoma had the highest number of species (species richness) to the lowest number of species (species richness) on tribal lands in Oklahoma using "Majority" data. The data listed below will be given from "Highest Majority" to "Lowest Majority." This data will show us what tribes specifically have the highest amount of species richness to the lowest amount of species richness on tribal-owned lands in Oklahoma. This data can then be used by biologists and conservation scientists to implement conservation efforts in conjunction with the tribes, if the need arises.

RANK-ID	TRIBES	HIGHEST MAJORITY (# OF SPECIES-SPECIES RICHNESS)
1	SENECA - CAYUGA TRIBE OF OKLAHOMA	123
2	CHEROKEE NATION, OKLAHOMA	123
3	MUSCOGEE (CREEK) NATION, OKLAHOMA	121
4	WYANDOTTE TRIBE OF OKLAHOMA	120
5	OTTAWA TRIBE OF INDIANS, OKLAHOMA	117
6	PEORIA TRIBE OF INDIANS OF OKLAHOMA	117
7	EASTERN SHAWNEE TRIBE OF OKLAHOMA	117
8	QUAPAW TRIBE OF INDIANS, OKLAHOMA	117
9	CHOCTAW NATION OF OKLAHOMA	114
10	CHEYENNE - ARAPAHO TRIBES OF OKLAHOMA	113
11	MIAMI TRIBE OF OKLAHOMA	112
12	MODOC TRIBE OF OKLAHOMA	111
13	KIOWA - COMANCHE - APACHE - FT. SILL APACHE	109
14	WICHITA - CADDO - DELAWARE	98
15	TONKAWA TRIBE OF INDIANS OF OKLAHOMA	95
16	PONCA TRIBE OF INDIANS OF OKLAHOMA	93
17	CHILOCCO INDAIN SCHOOL LANDS	85
18	KAW NATION, OKLAHOMA	84
19	SEMINOLE NATION OF OKLAHOMA	65
20	KICKAPOO TRIBE OF OKLAHOMA	63
21	POTAWATOMI - SHAWNEE NATIONS	60
22	CHICKASAW NATION, OKLAHOMA	58
23	PAWNEE NATION OF OKLAHOMA	58
24	OTOE - MISSOURI TRIBE OF INDIANS, OKLAHOMA	58
25	OSAGE TRIBE, OKLAHOMA	58
26	SAC AND FOX NATION, OKLAHOMA	57
27	IOWA TRIBE OF OKLAHOMA	57

Tyler Pounds-New Mexico State University-MA Anthropology (Ecological)

Now let us look at the data and the analyzed results. It should be noted that we used "Mean" data in order to determine what tribes have the most biodiversity on tribal-owned lands. The data will be listed from "Highest Mean" (most biodiversity) to "Lowest Mean" (least biodiversity). I am going to list the original data table below.

OBJECTID	TRIBAL_ARE	COUNT_	AREA	MIN_	MAX_	RANGE	MEAN	STD	SUM_	VARIETY	MAJORITY	MINORITY	MEDIAN
1	OTTAWA TRIBE OF INDIANS, OKLAHOMA	69176	62258400	24	168	144	116.8039	17.17582	8080025	124	117	54	118
2	WYANDOTTE TRIBE OF OKLAHOMA	95569	86012100	26	169	143	117.1499	17.60695	11195895	126	120	67	119
3	SEMINOLE NATION OF OKLAHOMA	1652079	1.49E+09	25	173	148	102.3801	30.82949	1.69E+08	146	65	30	116
4	WICHITA - CADDO - DELAWARE	3320218	2.99E+09	22	166	144	97.88982	23.20741	3.25E+08	142	98	22	99
5	CHICKASAW NATION, OKLAHOMA	21229785	1.91E+10	20	182	162	93.05466	29.55928	1.98E+09	163	58	21	101
6	CHILOCCO INDAIN SCHOOL LANDS	38205	34384500	32	165	133	96.55307	20.26152	3688810	112	85	35	90
7	CHEYENNE - ARAPAHO TRIBES OF OKLAHOMA	19467243	1.75E+10	0	167	167	103.9543	18.23664	2.02E+09	150	113	167	106
8	KIOWA - COMANCHE - APACHE - FT. SILL APACHE	13379511	1.2E+10	0	176	176	95.27787	23.15684	1.27E+09	159	109	176	100
9	PAWNEE NATION OF OKLAHOMA	1284112	1.16E+09	25	170	145	86.43124	30.60335	1.11E+08	145	58	170	67
10	PEORIA TRIBE OF INDIANS OF OKLAHOMA	108598	97738200	24	159	135	109.9457	15.32308	11939881	117	117	41	112
11	MIAMI TRIBE OF OKLAHOMA	116582	1.05E+08	24	167	143	107.953	19.72529	12585377	129	112	28	112
12	SENECA - CAYUGA TRIBE OF OKLAHOMA	234141	2.11E+08	22	159	137	108.6517	28.67657	25439827	128	123	56	117
13	TONKAWA TRIBE OF INDIANS OF OKLAHOMA	412909	3.72E+08	25	167	142	96.33118	18.4576	39776013	137	95	30	94
14	PONCA TRIBE OF INDIANS OF OKLAHOMA	470651	4.24E+08	24	176	152	94.37269	26.36619	44416603	151	93	28	95
15	EASTERN SHAWNEE TRIBE OF OKLAHOMA	61452	55306800	26	156	130	114.7326	14.03094	7050546	107	117	33	117
16	OTOE - MISSOURI TRIBE OF INDIANS, OKLAHOMA	580817	5.23E+08	22	171	149	75.74002	27.13076	43991091	147	58	32	61
17	MUSCOGEE (CREEK) NATION, OKLAHOMA	13989128	1.26E+10	16	180	164	105.9043	30.19446	1.48E+09	162	121	180	117
18	SAC AND FOX NATION, OKLAHOMA	2161664	1.95E+09	24	165	141	92.31711	32.10907	2E+08	139	57	32	97
19	IOWA TRIBE OF OKLAHOMA	1035944	9.32E+08	25	163	138	89.31103	30.26286	92521221	135	57	27	94
20	KICKAPOO TRIBE OF OKLAHOMA	922158	8.3E+08	25	166	141	92.56739	30.28596	85361760	140	63	32	101
21	POTAWATOMI - SHAWNEE NATIONS	2571018	2.31E+09	23	176	153	98.31924	30.68375	2.53E+08	150	60	176	113
22	CHOCTAW NATION OF OKLAHOMA	31395705	2.83E+10	16	183	167	109.2094	21.99849	3.43E+09	168	114	183	113
23	QUAPAW TRIBE OF INDIANS, OKLAHOMA	253442	2.28E+08	24	160	136	105.1928	20.55641	26660272	130	117	150	110
24	CHEROKEE NATION, OKLAHOMA	20015792	1.8E+10	16	184	168	111.156	25.8969	2.22E+09	166	123	183	118
25	OSAGE TRIBE, OKLAHOMA	6630598	5.97E+09	23	175	152	88.58196	29.5686	5.87E+08	152	58	175	86
26	KAW NATION, OKLAHOMA	448034	4.03E+08	23	148	125	95.53806	22.39327	42804300	124	84	31	98
27	MODOC TRIBE OF OKLAHOMA	21089	18980100	48	144	96	115.6614	10.2785	2439183	74	111	68	116

From the data that is listed below, we will be able to analyze what tribes in Oklahoma have the most biodiversity (Highest Mean) and the least biodiversity (Lowest Mean). This data can then be used by biologists and conservation scientists to implement conservation efforts in conjunction with the tribes, if the need arises.

RANK-ID	TRIBES	HIGHEST MEAN (MOST BIODIVERSITY)
1	WYANDOTTE TRIBE OF OKLAHOMA	117.1498603
2	OTTAWA TRIBE OF INDIANS, OKLAHOMA	116.8038771
3	MODOC TRIBE OF OKLAHOMA	115.6613875
4	EASTERN SHAWNEE TRIBE OF OKLAHOMA	114.7325718
5	CHEROKEE NATION, OKLAHOMA	111.1559642
6	PEORIA TRIBE OF INDIANS OF OKLAHOMA	109.9456804
7	CHOCTAW NATION OF OKLAHOMA	109.2093624
8	SENECA - CAYUGA TRIBE OF OKLAHOMA	108.6517398
9	MIAMI TRIBE OF OKLAHOMA	107.953003
10	MUSCOGEE (CREEK) NATION, OKLAHOMA	105.90428
11	QUAPAW TRIBE OF INDIANS, OKLAHOMA	105.1927936
12	CHEYENNE - ARAPAHO TRIBES OF OKLAHOMA	103.9542766
13	SEMINOLE NATION OF OKLAHOMA	102.3800708
14	POTAWATOMI - SHAWNEE NATIONS	98.31924242
15	WICHITA - CADDO - DELAWARE	97.8898244
16	CHILOCCO INDAIN SCHOOL LANDS	96.55306897
17	TONKAWA TRIBE OF INDIANS OF OKLAHOMA	96.33118435
18	KAW NATION, OKLAHOMA	95.53806184
19	KIOWA - COMANCHE - APACHE - FT. SILL APACHE	95.27786591
20	PONCA TRIBE OF INDIANS OF OKLAHOMA	94.37269442
21	CHICKASAW NATION, OKLAHOMA	93.05466287
22	KICKAPOO TRIBE OF OKLAHOMA	92.56739084
23	SAC AND FOX NATION, OKLAHOMA	92.31711219
24	IOWA TRIBE OF OKLAHOMA	89.3110255
25	OSAGE TRIBE, OKLAHOMA	88.58195897
26	PAWNEE NATION OF OKLAHOMA	86.43123653
27	OTOE - MISSOURI TRIBE OF INDIANS, OKLAHOMA	75.74001966

Tyler Pounds-New Mexico State University-MA Anthropology (Ecological)

Let's look at the original data table for the state of Oklahoma, as a whole. It should be noted that we are going to compare and contrast the data table for the State of Oklahoma area data with the tribal land data within Oklahoma.

OBJECTID	OKLAHOMA STATE_ARE	COUNT_	AREA	MIN_	MAX_	RANGE	MEAN	STD	SUM_	VARIETY	MAJORITY	MINORITY	MEDIAN
1	OKLAHOMA BOUNDARY SR	201454360	1.81309E+11	0	185	185	100.961664	24.960526	20339167311	173	113	184	105

Tyler Pounds-New Mexico State University-MA Anthropology (Ecological)

We are going to have to specifically compare and contrast the "Highest Majority" (Species Richness) and "Highest Mean" (Biodiversity) data that we analyzed above.

RANK-ID	TRIBES	HIGHEST MAJORITY (# OF SPECIES-SPECIES RICHNESS)
1	SENECA - CAYUGA TRIBE OF OKLAHOMA	123
2	CHEROKEE NATION, OKLAHOMA	123
3	MUSCOGEE (CREEK) NATION, OKLAHOMA	121
4	WYANDOTTE TRIBE OF OKLAHOMA	120
5	OTTAWA TRIBE OF INDIANS, OKLAHOMA	117
6	PEORIA TRIBE OF INDIANS OF OKLAHOMA	117
7	EASTERN SHAWNEE TRIBE OF OKLAHOMA	117
8	QUAPAW TRIBE OF INDIANS, OKLAHOMA	117
9	CHOCTAW NATION OF OKLAHOMA	114
10	CHEYENNE - ARAPAHO TRIBES OF OKLAHOMA	113
11	MIAMI TRIBE OF OKLAHOMA	112
12	MODOC TRIBE OF OKLAHOMA	111
13	KIOWA - COMANCHE - APACHE - FT. SILL APACHE	109
14	WICHITA - CADDO - DELAWARE	98
15	TONKAWA TRIBE OF INDIANS OF OKLAHOMA	95
16	PONCA TRIBE OF INDIANS OF OKLAHOMA	93
17	CHILOCCO INDAIN SCHOOL LANDS	85
18	KAW NATION, OKLAHOMA	84
19	SEMINOLE NATION OF OKLAHOMA	65
20	KICKAPOO TRIBE OF OKLAHOMA	63
21	POTAWATOMI - SHAWNEE NATIONS	60
22	CHICKASAW NATION, OKLAHOMA	58
23	PAWNEE NATION OF OKLAHOMA	58
24	OTOE - MISSOURI TRIBE OF INDIANS, OKLAHOMA	58
25	OSAGE TRIBE, OKLAHOMA	58
26	SAC AND FOX NATION, OKLAHOMA	57
27	IOWA TRIBE OF OKLAHOMA	57

Tyler Pounds-New Mexico State University-MA Anthropology (Ecological)

RANK-ID	TRIBES	HIGHEST MEAN (MOST BIODIVERSITY)
1	WYANDOTTE TRIBE OF OKLAHOMA	117.1498603
2	OTTAWA TRIBE OF INDIANS, OKLAHOMA	116.8038771
3	MODOC TRIBE OF OKLAHOMA	115.6613875
4	EASTERN SHAWNEE TRIBE OF OKLAHOMA	114.7325718
5	CHEROKEE NATION, OKLAHOMA	111.1559642
6	PEORIA TRIBE OF INDIANS OF OKLAHOMA	109.9456804
7	CHOCTAW NATION OF OKLAHOMA	109.2093624
8	SENECA - CAYUGA TRIBE OF OKLAHOMA	108.6517398
9	MIAMI TRIBE OF OKLAHOMA	107.953003
10	MUSCOGEE (CREEK) NATION, OKLAHOMA	105.90428
11	QUAPAW TRIBE OF INDIANS, OKLAHOMA	105.1927936
12	CHEYENNE - ARAPAHO TRIBES OF OKLAHOMA	103.9542766
13	SEMINOLE NATION OF OKLAHOMA	102.3800708
14	POTAWATOMI - SHAWNEE NATIONS	98.31924242
15	WICHITA - CADDO - DELAWARE	97.8898244
16	CHILOCCO INDAIN SCHOOL LANDS	96.55306897
17	TONKAWA TRIBE OF INDIANS OF OKLAHOMA	96.33118435
18	KAW NATION, OKLAHOMA	95.53806184
19	KIOWA - COMANCHE - APACHE - FT. SILL APACHE	95.27786591
20	PONCA TRIBE OF INDIANS OF OKLAHOMA	94.37269442
21	CHICKASAW NATION, OKLAHOMA	93.05466287
22	KICKAPOO TRIBE OF OKLAHOMA	92.56739084
23	SAC AND FOX NATION, OKLAHOMA	92.31711219
24	IOWA TRIBE OF OKLAHOMA	89.3110255
25	OSAGE TRIBE, OKLAHOMA	88.58195897
26	PAWNEE NATION OF OKLAHOMA	86.43123653
27	OTOE - MISSOURI TRIBE OF INDIANS, OKLAHOMA	75.74001966

Upon analyzing the data below, these are the tribes in Oklahoma whose "Highest Majority" (# of Species-Species Richness) is greater than the State of Oklahoma, combined.

RANK-ID TRIBES	HIGHEST MAJORITY (# OF SPECIES-SPECIES RICHNESS)
1 SENECA - CAYUGA TRIBE OF OKLAHOMA	123
2 CHEROKEE NATION, OKLAHOMA	123
3 MUSCOGEE (CREEK) NATION, OKLAHOMA	121
4 WYANDOTTE TRIBE OF OKLAHOMA	120
5 OTTAWA TRIBE OF INDIANS, OKLAHOMA	117
6 PEORIA TRIBE OF INDIANS OF OKLAHOMA	117
7 EASTERN SHAWNEE TRIBE OF OKLAHOMA	117
8 QUAPAW TRIBE OF INDIANS, OKLAHOMA	117
9 CHOCTAW NATION OF OKLAHOMA	114

Tyler Pounds-New Mexico State University-MA Anthropology (Ecological)

"Upon analyzing the data below, these are the tribes in Oklahoma whose "Highest Mean" (Most Biodiversity) is greater than the State of Oklahoma, combined.

RANK-ID TRIBES	HIGHEST MEAN (MOST BIODIVERSITY)
1 WYANDOTTE TRIBE OF OKLAHOMA	117.1498603
2 OTTAWA TRIBE OF INDIANS, OKLAHOMA	116.8038771
3 MODOC TRIBE OF OKLAHOMA	115.6613875
4 EASTERN SHAWNEE TRIBE OF OKLAHOMA	114.7325718
5 CHEROKEE NATION, OKLAHOMA	111.1559642
6 PEORIA TRIBE OF INDIANS OF OKLAHOMA	109.9456804
7 CHOCTAW NATION OF OKLAHOMA	109.2093624
8 SENECA - CAYUGA TRIBE OF OKLAHOMA	108.6517398
9 MIAMI TRIBE OF OKLAHOMA	107.953003
10 MUSCOGEE (CREEK) NATION, OKLAHOMA	105.90428
11 QUAPAW TRIBE OF INDIANS, OKLAHOMA	105.1927936
12 CHEYENNE - ARAPAHO TRIBES OF OKLAHOMA	103.9542766
13 SEMINOLE NATION OF OKLAHOMA	102.3800708

Tyler Pounds-New Mexico State University-MA Anthropology (Ecological)

Recommendations for Future Use of Data for Study

1. I recommend that the data in this Story Map be used in order to answer other various research questions directly related to the correlation between elevation in Oklahoma, the Highest Majority (# of Species-Species Richness), and the Highest Mean (Most Biodiversity).

2. I also recommend using the data listed above in the Story Map to look at the correlation between elevation in the state of Oklahoma and "Wildlife Habitat," in order to determine what areas of Oklahoma have higher wildlife populations than others.

GIS Project: Tribal Lands of Oklahoma
Analyzing Species Richness and Biodiversity on Tribal Lands
Tyler Pounds-New Mexico State University-MA Anthropology (Ecological)
December 10, 2019
https://storymaps.arcgis.com/stories/380e04df620d48c793f312751aedd188

As you can see above, GIS plays an integral role in the creation and integration of the BIOTEK Database Information Management System, the TRIBAL-RAD TEK System, and the GOV-RAD TEK System. It is my honor and duty to help humanity try to save as much biodiversity as is humanely and ethically possible. We must remember that humans are not invincible. The earth was here before humans ever were, and it will still be here if humans ever die out. We must be wise and live alongside it, without overusing resources or destroying biodiversity. If we do not follow this sacred teaching of the earth, humans will quite probably be destroyed forever. We have two paths on which to walk: we must choose the right path now.

We are now living in the Anthropocene Epoch (a human-caused geological time period of ecological destruction); it is imperative that we gather as much Traditional Ecological Knowledge (TEK) from Indigenous Peoples all over the world as possible. *Indigenous people and their traditional ecological knowledge base systems (TEK) are helping to preserve over 80 percent of the world's biodiversity Maryellen Kennedy Duckett, National Geographic, "Nature Needs us to Act-Now", March 4th, 2020.* I foresee my BIOTEK Databases being instrumental in documenting the biodiversity of plants and animals that Indigenous groups are managing on an international scale. This will also allow scientists to put "values" on the extent of Traditional Ecological Knowledge (TEK) bases of Indigenous groups all over the world. Documenting TEK through my BIOTEK Databases should also help to record changes in the field of "Global Climate Change."

Damian Carrington, *"The Anthropocene Epoch: Scientists Declare Dawn of Human-Influenced Age," The Guardian,* August 29, 2016, https://www.theguardian.com/environment/2016/aug/29/declare-anthropocene-epoch-experts-urge-geological-congress-human-impact-earth

National Geographic, *Voices for Biodiversity, "Traditional Ecological Knowledge, Anthropology and Climate Change,"* April 5, 2012, https://blog.nationalgeographic.org/2012/04/05/traditional-ecological-knowledge-anthropology-and-climate-change/

BIOTEK Database Information Management System Use in University/Research Settings

I eventually foresee universities, as well as conservation organizations all over the world, utilizing my BIOTEK Database Information Management System in order to document biodiversity and traditional knowledge bases amongst Indigenous Peoples. Note that the actual technology has yet to be developed and that, currently, I only have templates of how I would like to see my database information management system organized. Eventually, I plan on creating and designing a software program that uses these templates (as listed above). Once this software program has been created, universities and various conservation organizations all over the world will then be able to purchase my computer program in order to document biodiversity through a "Traditional Ecological Knowledge" framework. If any university or conservation organization would like to help me create/co-create my BIOTEK Database Information Management System into a cohesive software program, you can reach me at my e-mail address: tylerpounds1989@icloud.com.

Ecological Anthropology Becoming the Standard for Documenting "Uses of Biodiversity" (An Integrative Approach)

The formatting of documenting various uses for biodiversity through a cultural lens is depicted below. This can mainly be accomplished by utilizing "integrated ecosystem service approaches." This is how Ecological Anthropology should focus on documenting the uses of biodiversity for other scientists, as well as for their own records (as it will showcase "who, what, where, why,") and will answer fundamental questions regarding integrating Anthropology with Ecology, and other Biology-related fields. I use "Ethnographic Field Methods" in order to document plant usages. Other forms of studies could be utilized through database and archival retrievals, if necessary. When utilizing "Ethnographic Field Methods" to document the uses of plants and animals, it is also important to either take pictures of the specimens in question or to draw field sketches. This gives your peers and audience more physical and scientific evidence that such ecosystem services exist, and are currently being utilized within a "cultural, as well as ecological, framework."

Upon documenting uses of plants/animals utilizing ethnography and photographing/sketching specimens, the Ecological Anthropologist should then analyze the "Ethnographic Field Data," and then give as many charts, diagrams, and tables for this "Ethnographic Field Data" as possible, so that other scientists can interpret the "Ecosystem Services" data for themselves. One must also remember that Geographic Information Systems (GIS) is another invaluable tool for the fellow Ecological Anthropologist to utilize. GIS is useful for both Human Ecology, as well as Wildlife Ecology studies. Mapping species' habitats and populations of certain species in close proximity to people (or the culture utilizing the species in question) will be crucial for answering questions both ecological and cultural.

Many people should realize that Ecological Anthropology is a complementary field for documenting various uses of biodiversity, conserving wildlife, and natural resources. Ecological Anthropology is also a complementary, integrative scientific approach in explaining the links between human anthropogenic factors and the causes of "global climate change", destruction of biodiversity. After all, we are currently living in the "Anthropocene" geologic time period. Ecological Anthropology could also theoretically be utilized in explaining the link between wildlife species increasing or declining through human anthropogenic factors and cultural usages. This field of study utilizes both quantitative and qualitative approaches (which makes it an ideal form of documenting wildlife biodiversity from both a social science and biological science perspective).

Two other forms of anthropological "Field Methods" that I personally employ when conducting ethnobiological case studies in the field of Ecological Anthropology are utilizing voice recorders and video cameras. Utilizing "Visual Anthropology" is another way to captivate your audience, as well as capturing additional proof of "Ecosystem Service" species usages. In this way, the fields of Anthropology and Wildlife Ecology can be used to document ecological functions and uses for wildlife species. This also opens the door for Ecological Anthropology to be utilized within an "Integrative Biology" paradigm. The fusion of both qualitative and quantitative methodologies will have to be employed when trying to combat and solve problems and issues related to "global climate change" and "biodiversity mass extinction events." This is due to "global climate change" spanning many different social science and biological science fields simultaneously. What better way to solve issues related to "global climate change" and "biodiversity mass extinction events" than to use Ecological Anthropology and Integrative Biology during the "Anthropocene Epoch?"

The last thing that I wanted to mention is the usefulness of oral stories/oral histories when it comes to the relationship that people have with plants and animals. Many cultures have oral stories/oral histories with certain animal and plant species. Generally speaking, any culture that has a creation myth or story with any particular plant or animal species has a high probability for some form of "Ecosystem Service." Oral histories/creation stories can also provide useful insights into how a culture might be specifically utilizing the plant or animal species in question, or how likely the culture would be to either support or degrade the plant or animal species in question. For example, if a tribe of people considers cedar trees sacred in their usage for religious purposes, then it would be safe to say that the cedar tree population would likely be stable, or even increasing (as the need for cedar trees being used religiously would require it). In other words, oral histories and stories can provide clues or insights into what is happening to plant and animal species within any one particular culture.

Ecological Anthropology can also be utilized in the field of "Conservation Biology." My methods and systems of documenting species usage and biodiversity through ecosystem service approaches (while listing cultural usages and anthropogenic factors) will be critical in answering "who, what, where, why, and how" a species' population is either increasing or declining. We are living in the Anthropocene Epoch, so we know that humans are responsible for current species' destruction. The "who" would be the human culture that is utilizing the species in question for conservation, the "what" would be the plant or animal species needing to be conserved, the "where" would be the biogeographic region in which the conservation study is taking place, the "why" would be the underlying reason for the species either being supported or declining (various cultural usages and anthropogenic factors), and the "how" would be describing in-depth how the culture is either degrading or supporting a species' population, due to various anthropogenic factors.

So, if an anthropologist (such as myself) could figure out how to accurately depict "who, what, where, why, and how" a species' population is either increasing or decreasing due to various human anthropogenic factors, then it would be revolutionary for conserving the world's biodiversity. This would also open the doors for human-wildlife predictive modeling systems in order to somewhat predict what would happen to a wildlife species population (based off of what we know from a particular human culture and various anthropogenic factors that either support or degrade a species' population). This could help get a handle on the world's biodiversity crisis and prove essential for conserving both plant and animal species worldwide for the field of wildlife conservation. Together, we can save the planet. It is now time for us to open our minds to new possibilities and forms of science.

I do not believe that it is entirely possible for us as humans to entirely set aside land for wildlife uses anymore. This is mainly due to the fact that the world's human population is ever-increasing. As a result, as time goes on, the lands available to set aside for "wildlife-only" will diminish. This is why humans must figure out "how and why" plants and animals are important to them. Once this is figured out, then a successful plan for conservation through ecosystem services can be implemented.

We are living in a time where integrated sciences will become key in saving our global biodiversity and halting the "biodiversity mass extinction events" that are currently taking place. I also believe that humans and wildlife are directly intertwined with each other. This is because I am half-Shawnee (Native American) on my father's side and half-Celtic (Druid) on my mother's side, as both are earth-based belief systems that see themselves as part of nature. It is time for a change in how we view our relationship with nature, overall, as our fates are intertwined. This is the only way to make progress in the field of conservation, in my opinion.

Our relationship to nature is vital to spiritual and mental well-being, as well. These teachings are embedded within Indigenous cultures all around the world. In order to stay healthy, one must be balanced with the earth and oneself. The only real way to do that is to have a healthy relationship with nature. Where does most of our food come from? Nature.

Where do most of our crops grow? Outside in the fields amongst nature. Where do most of our minerals and wood products come from to build our houses? In the forests or mountains amongst nature. I think my readers now understand my point when I say that we as humans have to build better relationships with nature if we expect to continue to survive as a species on this planet (as natural resources are not infinite).

Solanum elaeagnifolium Cav.

Silverleaf Nightshade

Navajo Uses:

How to say "Silverleaf Nightshade" in Navajo= Ma'iid'a'a'

Drug, Eye Medicine

Use documented by:
Elmore, Francis H., 1944, Ethnobotany of the Navajo, Santa Fe, NM. School of American Research, page 75

Common names: Silverleaf Nightshade
Family: Solanaceae
Native American Tribe: Navajo
Use category: Drug
Use sub-category: Eye Medicine
Notes: Plant used for sore eyes.

Drug, Nose Medicine

Use documented by:
Elmore, Francis H., 1944, Ethnobotany of the Navajo, Santa Fe, NM. School of American Research, page 97

Common names: Silverleaf Nightshade
Family: Solanaceae
Native American Tribe: Navajo
Use category: Drug
Use sub-category: Nose Medicine
Notes: Plant used for nose troubles.

Drug, Throat Aid

Use documented by:
Elmore, Francis H., 1944, Ethnobotany of the Navajo, Santa Fe, NM. School of American Research, page 97

Common names: Silverleaf Nightshade
Family: Solanaceae
Native American Tribe: Navajo
Use category: Drug
Use sub-category: Throat Aid
Notes: Plant used for throat troubles.

Food, Cooking Agent

Use documented by:
Steggerda, Morris, 1941, Navajo Foods and their Preparation, Journal of the American Dietetic Association 17(3):217-25, page 222

Common names: Silverleaf Nightshade
Family: Solanaceae
Native American Tribe: Navajo
Use category: Food
Use sub-category: Cooking Agent
Notes: Dried or fresh berries added to goat's milk to make it curdle for cheese.

Zuni Uses:

Drug, Snake Bite Remedy

Use documented by:
Camazine, Scott and Robert A. Bye, 1980, A Study of the Medical Ethnobotany of the Zuni Indians of New Mexico, Journal of Ethnopharmacology 2:365-388, page 378

Common names: Silverleaf Nightshade
Family: Solanaceae
Native American Tribe: Zuni
Use category: Drug
Use sub-category: Snake Bite Remedy
Notes: Fresh or dried root chewed by medicine man before sucking snakebite and poultice applied to wound.

Drug, Toothache Remedy

Use documented by:
Stevenson, Matilda Coxe, 1915, Ethnobotany of the Zuni Indians, SI-BAE
Annual Report #30, page 60

Common names: Silverleaf Nightshade
Family: Solanaceae
Native American Tribe: Zuni
Use category: Drug
Use sub-category: Toothache Remedy
Notes: Chewed root placed in cavity of aching tooth.

Drug, Toothache Remedy

Use documented by:
Camazine, Scott and Robert A. Bye, 1980, A Study of the Medical
Ethnobotany of the Zuni Indians of New Mexico, Journal of
Ethnopharmacology 2:365-388, page 378

Common names: Silverleaf Nightshade
Family: Solanaceae
Native American Tribe: Zuni
Use category: Drug
Use sub-category: Toothache Remedy
Notes: Fruit chewed over sore tooth.

Food, Beverage

Use documented by:
Stevenson, Matilda Coxe, 1915, Ethnobotany of the Zuni Indians, SI-BAE
Annual Report #30, page 70

Common names: Silverleaf Nightshade
Family: Solanaceae
Native American Tribe: Zuni

Use category: Food
Use sub-category: Beverage
Notes: Berries mixed with curdled goat milk and considered a delicious beverage.

Cochiti-Pueblo Uses:

Food, Substitution Food

Use documented by:
Castetter, Edward F., 1935, Ethnobiological Studies in the American Southwest I. Uncultivated Native Plants Used as Sources of Food, University of New Mexico Bulletin 4(1):1-44, page 51

Common names: Silverleaf Nightshade
Family: Solanaceae
Native American Tribe: Cochiti
Use category: Food
Use sub-category: Substitution Food
Notes: Fruits used as a substitute for rennet in curdling milk.

Isleta-Pueblo Uses:

Drug, Laxative

Use documented by:
Jones, Volney H., 1931, The Ethnobotany of the Isleta Indians, University of New Mexico, M.A. Thesis, page 43

Common names: Silverleaf Nightshade
Family: Solanaceae
Native American Tribe: Isleta
Use category: Drug
Use sub-category: Laxative
Notes: Raw seed pods eaten or boiled into a syrup and taken as a laxative.

Drug, Gynecological Aid

Use documented by:
**White, Leslie A., 1945, Notes on the Ethnobotany of the Keres, Papers of the
Michigan Academy of Arts, Sciences, and Letters 30:557-568, page 562**

Common names: Silverleaf Nightshade
Family: Solanaceae
Native American Tribe: Keresan
Use category: Drug
Use sub-category: Gynecological Aid
Notes: Infusion of plant taken by nursing mothers to sustain milk flow.

Other, Jewelry

Use documented by:
**White, Leslie A., 1945, Notes on the Ethnobotany of the Keres, Papers of the
Michigan Academy of Arts, Sciences, and Letters 30:557-568, page 562**

Common names: Silverleaf Nightshade
Family: Solanaceae
Native American Tribe: Keresan
Use category: Other
Use sub-category: Jewelry
Notes: Fruit made into a necklace worn by women.

Other, Jewelry

Use documented by:
**Whiting, Alfred F., 1939, Ethnobotany of the Hopi, Museum of Northern
Arizona Bulletin #15, page 90**

Common names: Silverleaf Nightshade
Family: Solanaceae
Native American Tribe: Hopi
Use category: Other
Use sub-category: Jewelry
Notes: Yellow fruits made into necklaces for clowns.

Pima Uses:

Drug, Cold Remedy

Use documented by:
Curtin, L. S. M., 1949, By the Prophet of the Earth, Santa Fe. San Vicente Foundation, page 88

Common names: Silverleaf Nightshade
Family: Solanaceae
Native American Tribe: Pima
Use category: Drug
Use sub-category: Cold Remedy
Notes: Crushed, dried berries used for colds.

Food, Fruit

Use documented by:
Curtin, L. S. M., 1949, By the Prophet of the Earth, Santa Fe. San Vicente Foundation, page 88

Common names: Silverleaf Nightshade
Family: Solanaceae
Native American Tribe: Pima
Use category: Food
Use sub-category: Fruit
Notes: Berries powdered, placed in milk, a piece of rabbit or cow stomach added, and liquid eaten as cheese.

Food, Substitution Food

**Use documented by:
Russell, Frank, 1908, The Pima Indians, SI-BAE Annual Report #26:1-390,
page 78**

Common names: Silverleaf Nightshade
Family: Solanaceae
Native American Tribe: Pima
Use category: Food
Use sub-category: Substitution Food
Notes: Berries used as a substitute for rennet.

Apache-White Mountain Uses:

Drug, Unspecified

**Use documented by:
Reagan, Albert B., 1929, Plants Used by the White Mountain Apache
Indians of Arizona, Wisconsin Archeologist 8:143-61, page 160**

Common names: Silverleaf Nightshade
Family: Solanaceae
Native American Tribe: Apache, White Mountain
Use category: Drug
Use sub-category: Unspecified
Notes: Plant used for medicinal purposes.

Additional Cultural Usages:

Spanish-American Uses:

Food, Substitution Food

Use documented by:
Castetter, Edward F., 1935, Ethnobiological Studies in the American
Southwest I. Uncultivated Native Plants Used as Sources of Food, University
of New Mexico Bulletin 4(1):1-44, page 51

Common names: Silverleaf Nightshade
Family: Solanaceae
Native American Tribe: Spanish American
Use category: Food
Use sub-category: Substitution Food
Notes: Fruits used as a substitute for rennet in curdling milk.

Populus deltoides wislizenii

Rio Grande Cottonwood

Food, Candy

Use documented by:
Castetter, Edward F., 1935, Ethnobiological Studies in the American
Southwest I. Uncultivated Native Plants Used as Sources of Food, University
of New Mexico Bulletin 4(1):1-44, page 31

Common names: Rio Grande Cottonwood
Family: Salicaceae
Native American Tribe: Acoma
Use category: Food
Use sub-category: Candy
Notes: Cotton from the pistillate catkins used as chewing gum.

Food, Candy

Use documented by:
Castetter, Edward F. and M. E. Opler, 1936, Ethnobiological Studies in the American Southwest III. The Ethnobiology of the Chiricahua and Mescalero Apache, University of New Mexico Bulletin 4(5):1-63, page 45

Common names: Rio Grande Cottonwood
Family: Salicaceae
Native American Tribe: Apache, Chiricahua and Mescalero
Use category: Food
Use sub-category: Candy
Notes: Buds used as chewing gum.

Food, Candy

Use documented by:
Reagan, Albert B., 1929, Plants Used by the White Mountain Apache Indians of Arizona, Wisconsin Archeologist 8:143-61, page 159

Common names: Rio Grande Cottonwood
Family: Salicaceae
Native American Tribe: Apache, White Mountain
Use category: Food
Use sub-category: Candy
Notes: Buds used as chewing gum.

Fiber, Building Material

Use documented by:
Jones, Volney H., 1931, The Ethnobotany of the Isleta Indians, University of New Mexico, M.A. Thesis, page 39

Common names: Rio Grande Cottonwood
Family: Salicaceae
Native American Tribe: Isleta
Use category: Fiber
Use sub-category: Building Material
Notes: Smaller limbs and leaves used for thatching houses.

Fiber, Canoe Material

Use documented by:
Jones, Volney H., 1931, The Ethnobotany of the Isleta Indians, University of New Mexico, M.A. Thesis, page 39

Common names: Rio Grande Cottonwood
Family: Salicaceae
Native American Tribe: Isleta
Use category: Fiber
Use sub-category: Canoe Material
Notes: Wood formerly used in making small boats and rafts.

Food, Candy

Use documented by:
Jones, Volney H., 1931, The Ethnobotany of the Isleta Indians, University of New Mexico, M.A. Thesis, page 39

Common names: Rio Grande Cottonwood
Family: Salicaceae

Native American Tribe: Isleta
Use category: Food
Use sub-category: Candy
Notes: Fruit used by children for chewing gum.

Other, Cash Crop

Use documented by:
Jones, Volney H., 1931, The Ethnobotany of the Isleta Indians, University of New Mexico, M.A. Thesis, page 39

Common names: Rio Grande Cottonwood
Family: Salicaceae
Native American Tribe: Isleta
Use category: Other
Use sub-category: Cash Crop
Notes: Limbs used to make small bows and arrows for sale to tourists.

Food, Unspecified

Use documented by:
Castetter, Edward F., 1935, Ethnobiological Studies in the American Southwest I. Uncultivated Native Plants Used as Sources of Food, University of New Mexico Bulletin 4(1):1-44, page 43

Common names: Rio Grande Cottonwood
Family: Salicaceae
Native American Tribe: Isleta
Use category: Food
Use sub-category: Unspecified
Notes: Catkins eaten raw.

Jemez-Pueblo Uses:

Food, Unspecified

Use documented by:
Castetter, Edward F., 1935, Ethnobiological Studies in the American
Southwest I. Uncultivated Native Plants Used as Sources of Food, University
of New Mexico Bulletin 4(1):1-44, page 43

Common names: Rio Grande Cottonwood
Family: Salicaceae
Native American Tribe: Jemez
Use category: Food
Use sub-category: Unspecified
Notes: Catkins eaten raw.

Keres (Western)-Pueblo Uses:

Food, Candy

Use documented by:
Swank, George R., 1932, The Ethnobotany of the Acoma and Laguna
Indians, University of New Mexico, M.A. Thesis, page 62

Common names: Rio Grande Cottonwood
Family: Salicaceae
Native American Tribe: Keres, Western
Use category: Food
Use sub-category: Candy
Notes: Cotton used by children for chewing gum.

Other, Ceremonial Items

Use documented by:
Swank, George R., 1932, The Ethnobotany of the Acoma and Laguna
Indians, University of New Mexico, M.A. Thesis, page 62

Common names: Rio Grande Cottonwood
Family: Salicaceae
Native American Tribe: Keres, Western
Use category: Other
Use sub-category: Ceremonial Items
Notes: Twigs mixed with spruce branches the day after the mask dance.

Other, Fuel

Use documented by:
Swank, George R., 1932, The Ethnobotany of the Acoma and Laguna Indians, University of New Mexico, M.A. Thesis, page 62

Common names: Rio Grande Cottonwood
Family: Salicaceae
Native American Tribe: Keres, Western
Use category: Other
Use sub-category: Fuel
Notes: Wood used for fuel.

Laguna-Pueblo Uses:

Food, Candy

Use documented by:
Castetter, Edward F., 1935, Ethnobiological Studies in the American Southwest I. Uncultivated Native Plants Used as Sources of Food, University of New Mexico Bulletin 4(1):1-44, page 31

Common names: Rio Grande Cottonwood
Family: Salicaceae
Native American Tribe: Laguna
Use category: Food
Use sub-category: Candy
Notes: Cotton from the pistillate catkins used as chewing gum.

Fiber, Building Material

Use documented by:
Elmore, Francis H., 1944, Ethnobotany of the Navajo, Santa Fe, NM. School
of American Research, page 38

Common names: Rio Grande Cottonwood
Family: Salicaceae
Native American Tribe: Navajo
Use category: Fiber
Use sub-category: Building Material
Notes: Wood used for firewood, fence posts, vigas (heavy rafters), and tinder
boxes.

Fiber, Furniture

Use documented by:
Elmore, Francis H., 1944, Ethnobotany of the Navajo, Santa Fe, NM. School
of American Research, page 38

Common names: Rio Grande Cottonwood
Family: Salicaceae
Native American Tribe: Navajo
Use category: Fiber
Use sub-category: Furniture
Notes: Wood used to make cradles.

Food, Candy

Use documented by:
Reagan, Albert B., 1929, Plants Used by the White Mountain Apache
Indians of Arizona, Wisconsin Archeologist 8:143-61, page 159

Common names: Rio Grande Cottonwood
Family: Salicaceae
Native American Tribe: Navajo
Use category: Food
Use sub-category: Candy
Notes: Buds used as chewing gum.

Food, Candy

Use documented by:
**Elmore, Francis H., 1944, Ethnobotany of the Navajo, Santa Fe, NM. School
of American Research, page 38**

Common names: Rio Grande Cottonwood
Family: Salicaceae
Native American Tribe: Navajo
Use category: Food
Use sub-category: Candy
Notes: Sap or catkins, alone or mixed with animal fat, used for chewing gum.

Other, Ceremonial Items

Use documented by:
**Elmore, Francis H., 1944, Ethnobotany of the Navajo, Santa Fe, NM. School
of American Research, page 38**

Common names: Rio Grande Cottonwood
Family: Salicaceae
Native American Tribe: Navajo
Use category: Other
Use sub-category: Ceremonial Items
Notes: Wood used to carve dolls and images of some animals for ceremonial
purposes.

Other, Containers

**Use documented by:
Elmore, Francis H., 1944, Ethnobotany of the Navajo, Santa Fe, NM. School
of American Research, page 38**

Common names: Rio Grande Cottonwood
Family: Salicaceae
Native American Tribe: Navajo
Use category: Other
Use sub-category: Containers
Notes: Wood used to make tinder boxes.

Other, Fuel

**Use documented by:
Elmore, Francis H., 1944, Ethnobotany of the Navajo, Santa Fe, NM. School
of American Research, page 38**

Common names: Rio Grande Cottonwood
Family: Salicaceae
Native American Tribe: Navajo
Use category: Other
Use sub-category: Fuel
Notes: Wood used for firewood.

Other, Tools

**Use documented by:
Elmore, Francis H., 1944, Ethnobotany of the Navajo, Santa Fe, NM. School
of American Research, page 38**

Common names: Rio Grande Cottonwood
Family: Salicaceae
Native American Tribe: Navajo
Use category: Other

Use sub-category: Tools
Notes: Used to make wooden tubes for the bellows used in silversmithing.

Pima Uses:

Food, Candy

Use documented by:
Hrdlicka, Ales, 1908, Physiological and Medical Observations Among the Indians of Southwestern United States and Northern Mexico, SI-BAE Bulletin #34:1-427, page 265

Common names: Rio Grande Cottonwood
Family: Salicaceae
Native American Tribe: Pima
Use category: Food
Use sub-category: Candy
Notes: Buds used for chewing gum in early spring.

Tewa Uses:

Other, Decorations

Use documented by:
Robbins, W.W., J.P. Harrington, and B. Freire-Marreco, 1916, Ethnobotany of the Tewa Indians, SI-BAE Bulletin #55, page 42

Common names: Rio Grande Cottonwood
Family: Salicaceae
Native American Tribe: Tewa
Use category: Other
Use sub-category: Decorations
Notes: Wood used to make many artifacts.

Zuni Uses:

Food, Candy

Use documented by:
Reagan, Albert B., 1929, Plants Used by the White Mountain Apache
Indians of Arizona, Wisconsin Archeologist 8:143-61, page 159

Common names: Rio Grande Cottonwood
Family: Salicaceae
Native American Tribe: Zuni
Use category: Food
Use sub-category: Candy
Notes: Buds used as chewing gum.

Helianthus petiolaris
Prairie Sunflower

Food, Dried Food

Use documented by:
Weber, Steven A. and P. David Seaman, 1985, Havasupai Habitat: A. F.
Whiting's Ethnography of a Traditional Indian Culture, Tucson. The
University of Arizona Press, page 248

Common names: Prairie Sunflower
Family: Asteraceae
Native American Tribe: Havasupai
Use category: Food
Use sub-category: Dried Food
Notes: Seeds sun-dried and stored for winter use.

Food, Preserves

Use documented by:
Weber, Steven A. and P. David Seaman, 1985, Havasupai Habitat: A. F.
Whiting's Ethnography of a Traditional Indian Culture, Tucson. The
University of Arizona Press, page 67

Common names: Prairie Sunflower
Family: Asteraceae
Native American Tribe: Havasupai
Use category: Food
Use sub-category: Preserves
Notes: Seeds parched, ground, kneaded into seed butter, and eaten with fruit
drinks, or spread on bread.

Food, Staple

Use documented by:
Weber, Steven A. and P. David Seaman, 1985, Havasupai Habitat: A. F.
Whiting's Ethnography of a Traditional Indian Culture, Tucson. The
University of Arizona Press, page 67

Common names: Prairie Sunflower
Family: Asteraceae
Native American Tribe: Havasupai
Use category: Food
Use sub-category: Staple
Notes: Seeds ground and eaten as a ground or parched meal.

Hopi Uses:

Drug, Dermatological Aid

Use documented by:
Whiting, Alfred F., 1939, Ethnobotany of the Hopi, Museum of Northern
Arizona Bulletin #15, page 32, 96

Common names: Prairie Sunflower
Family: Asteraceae
Native American Tribe: Hopi
Use category: Drug
Use sub-category: Dermatological Aid
Notes: Plant used as a 'spider bite medicine.'

Food, Fodder

Use documented by:
Whiting, Alfred F., 1939, Ethnobotany of the Hopi, Museum of Northern
Arizona Bulletin #15, page 96

Common names: Prairie Sunflower
Family: Asteraceae
Native American Tribe: Hopi
Use category: Food
Use sub-category: Fodder
Notes: Used as an important food for summer birds.

Other, Ceremonial Items

Use documented by:
Colton, Harold S., 1974, Hopi History and Ethnobotany, IN D. A. Horr (ed.)
Hopi Indians. Garland: New York, page 324

Common names: Prairie Sunflower
Family: Asteraceae
Native American Tribe: Hopi
Use category: Other
Use sub-category: Ceremonial Items
Notes: Dried petals ground and mixed with corn meal to make yellow face
powder for women's basket dance.

Other, Decorations

Use documented by:
Colton, Harold S., 1974, Hopi History and Ethnobotany, IN D. A. Horr (ed.)
Hopi Indians. Garland: New York, page 324

Common names: Prairie Sunflower
Family: Asteraceae
Native American Tribe: Hopi
Use category: Other
Use sub-category: Decorations
Notes: Whole plant used in the decoration of flute priests in the Flute Ceremony.

Other, Season Indicator

Use documented by:
Colton, Harold S., 1974, Hopi History and Ethnobotany, IN D. A. Horr (ed.)
Hopi Indians. Garland: New York, page 324

Common names: Prairie Sunflower
Family: Asteraceae
Native American Tribe: Hopi
Use category: Other
Use sub-category: Season Indicator
Notes: Amount of flowers present used as a sign that there will be copious rains
and abundant harvest.

Navajo Uses:
*How to say "Prairie Sunflower" in Navajo= ndiyi'liitsoh (Sunflower; closest
translation)*

Drug, Hunting Medicine

Use documented by:
Vestal, Paul A., 1952, The Ethnobotany of the Ramah Navaho, Papers of the
Peabody Museum of American Archaeology and Ethnology 40(4):1-94, page
52

Common names: Prairie Sunflower
Family: Asteraceae
Native American Tribe: Navajo, Ramah
Use category: Drug
Use sub-category: Hunting Medicine
Notes: Cold infusion of flowers sprinkled on clothing for good luck in hunting.

Drug, Panacea

Use documented by:
Vestal, Paul A., 1952, The Ethnobotany of the Ramah Navaho, Papers of the Peabody Museum of American Archaeology and Ethnology 40(4):1-94, page 52

Common names: Prairie Sunflower
Family: Asteraceae
Native American Tribe: Navajo, Ramah
Use category: Drug
Use sub-category: Panacea
Notes: Cold infusion of whole plant used as 'life medicine.'

Chloracantha spinosa
Mexican Devil-Weed

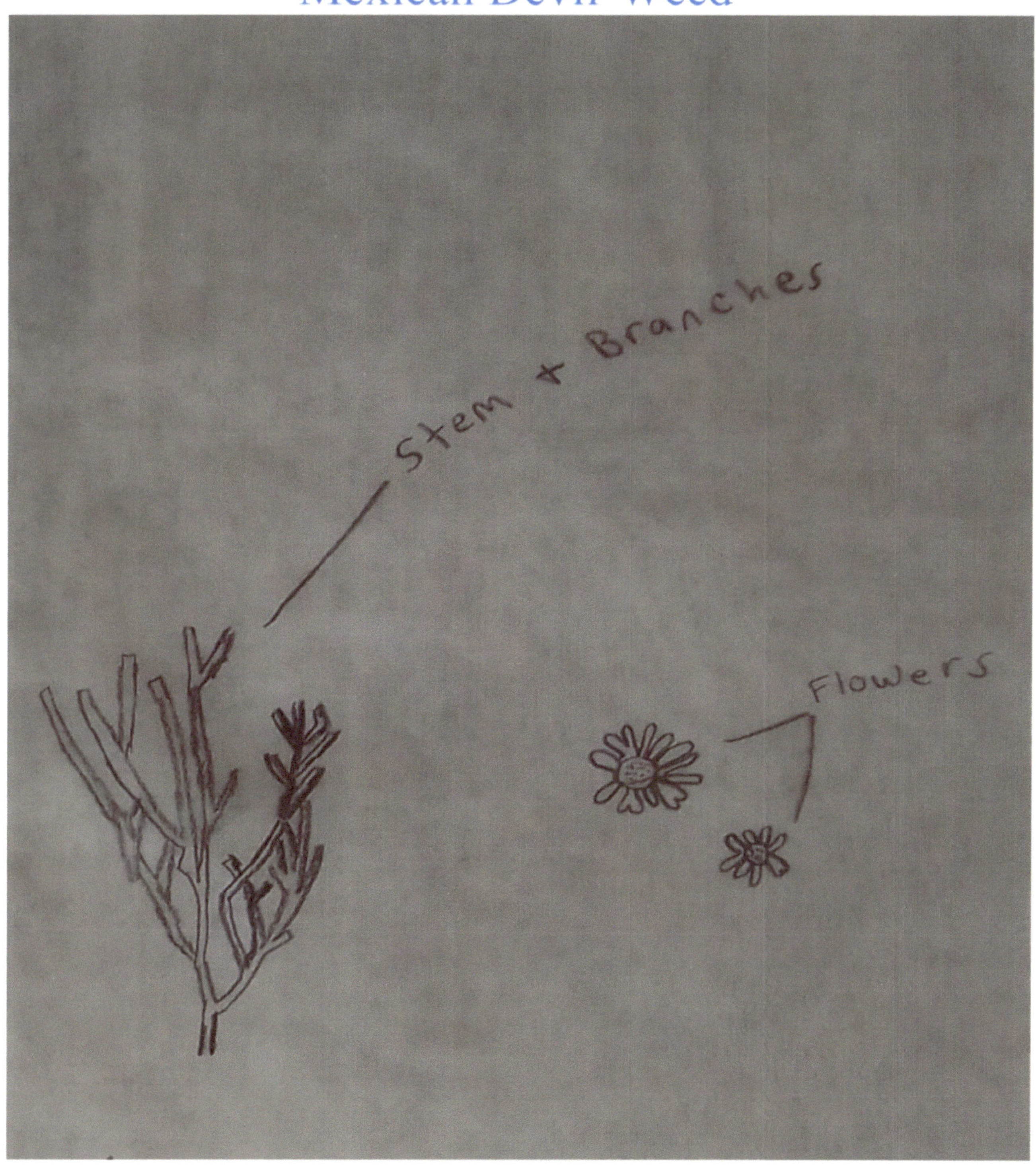

Food, Starvation Food

Use documented by:
Castetter, Edward F. and Willis H. Bell, 1951, Yuman Indian Agriculture, Albuquerque: University of New Mexico Press, page 201

Common names: Mexican Devil-Weed
Family: Asteraceae
Native American Tribe: Mojave
Use category: Food
Use sub-category: Starvation Food
Notes: Young shoots roasted and eaten as a famine food.

Food, Candy

Elmore, Francis H., 1944, Ethnobotany of the Navajo, Santa Fe, NM. School of American Research, page 83

Common names: Mexican Devil-Weed
Family: Asteraceae
Native American Tribe: Navajo
Use category: Food
Use sub-category: Candy
Notes: Stems chewed for gum.

Anemopsis californica
Yerba Mansa

Food, Bread and Cake

Use documented by:
Gifford, E. W., 1931, The Kamia of Imperial Valley, Washington, D.C., U.S.
Government Printing Office, page 24

Common names: Yerba Mansa
Family: Saururaceae
Native American Tribe: Kamia
Use category: Food
Use sub-category: Bread and Cake
Notes: Pulverized seeds used for bread.

Food, Porridge

Use documented by:
Gifford, E. W., 1931, The Kamia of Imperial Valley, Washington, D.C., U.S.
Government Printing Office, page 24

Common names: Yerba Mansa
Family: Saururaceae
Native American Tribe: Kamia
Use category: Food
Use sub-category: Porridge
Notes: Pulverized seeds cooked as mush.

Drug, Burn Dressing

Use documented by:
Swank, George R., 1932, The Ethnobotany of the Acoma and Laguna Indians, University of New Mexico, M.A. Thesis, page 26

Common names: Yerba Mansa
Family: Saururaceae
Native American Tribe: Keres, Western
Use category: Drug
Use sub-category: Burn Dressing
Notes: Poultice of green, chewed leaves applied to burns.

Drug, Dermatological Aid

Use documented by:
Swank, George R., 1932, The Ethnobotany of the Acoma and Laguna Indians, University of New Mexico, M.A. Thesis, page 26

Common names: Yerba Mansa
Family: Saururaceae
Native American Tribe: Keres, Western
Use category: Drug
Use sub-category: Dermatological Aid
Notes: Dried leaves ground into a powder and used on open sores.

Drug, Laxative

Use documented by:
Steward, Julian H., 1933, Ethnography of the Owens Valley Paiute,
University of California Publications in American Archaeology and
Ethnology 33(3):233-250, page 317

Common names: Yerba Mansa
Family: Saururaceae
Native American Tribe: Paiute
Use category: Drug
Use sub-category: Laxative
Notes: Infusion of roots taken as a laxative.

Drug, Orthopedic Aid

Use documented by:
Train, Percy, James R. Henrichs, and W. Andrew Archer, 1941, Medicinal
Uses of Plants by Indian Tribes of Nevada, Washington DC., U.S.
Department of Agriculture, page 33, 34

Common names: Yerba Mansa
Family: Saururaceae
Native American Tribe: Paiute
Use category: Drug
Use sub-category: Orthopedic Aid
Notes: Decoction of leaves used as a bath for muscular pains and sore feet.

Drug, Venereal Aid

Use documented by:
Steward, Julian H., 1933, Ethnography of the Owens Valley Paiute,
University of California Publications in American Archaeology and
Ethnology 33(3):233-250, page 317

Common names: Yerba Mansa
Family: Saururaceae
Native American Tribe: Paiute
Use category: Drug
Use sub-category: Venereal Aid
Notes: Infusion of roots taken for gonorrhea.

Papago Uses:

Drug, Emetic

Use documented by:
**Castetter, Edward F. and Ruth M. Underhill, 1935, Ethnobiological Studies
in the American Southwest II. The Ethnobiology of the Papago Indians,
University of New Mexico Bulletin 4(3):1-84, page 65**

Common names: Yerba Mansa
Family: Saururaceae
Native American Tribe: Papago
Use category: Drug
Use sub-category: Emetic
Notes: Decoction of leaves taken as an emetic.

Pima Uses:

Drug, Cold Remedy

Use documented by:
**Curtin, L. S. M., 1949, By the Prophet of the Earth, Santa Fe. San Vicente
Foundation, page 78**

Common names: Yerba Mansa
Family: Saururaceae
Native American Tribe: Pima
Use category: Drug

Use sub-category: Cold Remedy
Notes: Infusion of dried roots or plant taken for colds.

Drug, Cough Medicine

**Use documented by:
Curtin, L. S. M., 1949, By the Prophet of the Earth, Santa Fe. San Vicente
Foundation, page 78**

Common names: Yerba Mansa
Family: Saururaceae
Native American Tribe: Pima
Use category: Drug
Use sub-category: Cough Medicine
Notes: Roots chewed and swallowed, or decoction of roots taken for coughs.

Drug, Dermatological Aid

**Use documented by:
Curtin, L. S. M., 1949, By the Prophet of the Earth, Santa Fe. San Vicente
Foundation, page 78**

Common names: Yerba Mansa
Family: Saururaceae
Native American Tribe: Pima
Use category: Drug
Use sub-category: Dermatological Aid
Notes: Decoction of plant used as a wash, and poultice of leaves applied to
wounds.

Drug, Diaphoretic

**Use documented by:
Curtin, L. S. M., 1949, By the Prophet of the Earth, Santa Fe. San Vicente
Foundation, page 78**

Common names: Yerba Mansa
Family: Saururaceae
Native American Tribe: Pima
Use category: Drug
Use sub-category: Diaphoretic
Notes: Infusion of plant taken for colds, and to cause sweating.

Drug, Emetic

Use documented by:
Russell, Frank, 1908, The Pima Indians, SI-BAE Annual Report #26:1-390,
page 80

Common names: Yerba Mansa
Family: Saururaceae
Native American Tribe: Pima
Use category: Drug
Use sub-category: Emetic
Notes: Decoction of crushed root taken as an emetic.

Drug, Gastrointestinal Aid

Use documented by:
Curtin, L. S. M., 1949, By the Prophet of the Earth, Santa Fe. San Vicente
Foundation, page 78

Common names: Yerba Mansa
Family: Saururaceae
Native American Tribe: Pima
Use category: Drug
Use sub-category: Gastrointestinal Aid
Notes: Poultice of wet, powdered roots applied for stomachaches.

Drug, Other

Use documented by:
Curtin, L. S. M., 1949, By the Prophet of the Earth, Santa Fe. San Vicente
Foundation, page 78

Common names: Yerba Mansa
Family: Saururaceae
Native American Tribe: Pima
Use category: Drug
Use sub-category: Other
Notes: Infusion of roots taken and used as a wash for 'bad disease.'

Drug, Throat Aid

Use documented by:
Curtin, L. S. M., 1949, By the Prophet of the Earth, Santa Fe. San Vicente
Foundation, page 78

Common names: Yerba Mansa
Family: Saururaceae
Native American Tribe: Pima
Use category: Drug
Use sub-category: Throat Aid
Notes: Dry root held in the mouth for sore throat, and infusion taken for itchy
throat.

Drug, Tuberculosis Remedy

Use documented by:
Russell, Frank, 1908, The Pima Indians, SI-BAE Annual Report #26:1-390,
page 80

Common names: Yerba Mansa
Family: Saururaceae
Native American Tribe: Pima

Use category: Drug
Use sub-category: Tuberculosis Remedy
Notes: Decoction of crushed root taken for consumption.

Drug, Venereal Aid

**Use documented by:
Curtin, L. S. M., 1949, By the Prophet of the Earth, Santa Fe. San Vicente
Foundation, page 78**

Common names: Yerba Mansa
Family: Saururaceae
Native American Tribe: Pima
Use category: Drug
Use sub-category: Venereal Aid
Notes: Infusion of roots used as a wash for syphilis.

Prosopis pubescens
Screwbean Mesquite

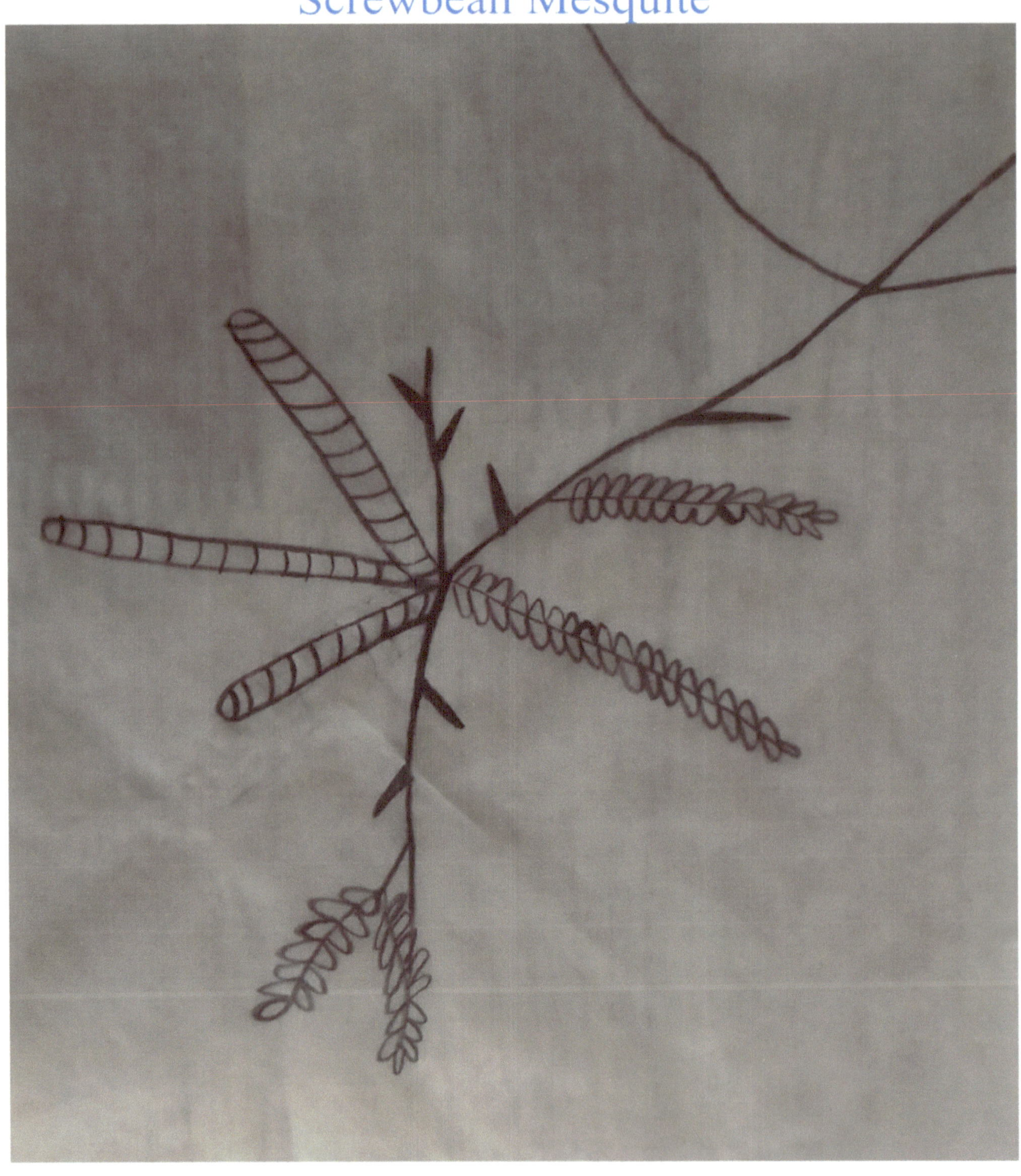

Apache (Chiricahua/Mescalero) Uses:

Food, Beverage

Use documented by:
**Castetter, Edward F. and M. E. Opler, 1936, Ethnobiological Studies in the
American Southwest III. The Ethnobiology of the Chiricahua and Mescalero
Apache, University of New Mexico Bulletin 4(5):1-63, page 53**

Common names: Screwbean Mesquite
Family: Fabaceae
Native American Tribe: Apache, Chiricahua and Mescalero
Use category: Food
Use sub-category: Beverage
Notes: Fruit ground and sugar added to make a thick drink.

Food, Bread and Cake

Use documented by:
**Castetter, Edward F. and M. E. Opler, 1936, Ethnobiological Studies in the
American Southwest III. The Ethnobiology of the Chiricahua and Mescalero
Apache, University of New Mexico Bulletin 4(5):1-63, page 41**

Common names: Screwbean Mesquite
Family: Fabaceae
Native American Tribe: Apache, Chiricahua and Mescalero
Use category: Food
Use sub-category: Bread and Cake
Notes: Pods dried, washed, ground into flour, and made into bread.

Food, Dried Food

Use documented by:
**Castetter, Edward F. and M. E. Opler, 1936, Ethnobiological Studies in the
American Southwest III. The Ethnobiology of the Chiricahua and Mescalero
Apache, University of New Mexico Bulletin 4(5):1-63, page 41**

Common names: Screwbean Mesquite
Family: Fabaceae
Native American Tribe: Apache, Chiricahua and Mescalero
Use category: Food
Use sub-category: Dried Food
Notes: Fruits gathered, dried, and stored in sacks.

Food, Special Food

Use documented by:
Castetter, Edward F. and M. E. Opler, 1936, Ethnobiological Studies in the American Southwest III. The Ethnobiology of the Chiricahua and Mescalero Apache, University of New Mexico Bulletin 4(5):1-63, page 41

Common names: Screwbean Mesquite
Family: Fabaceae
Native American Tribe: Apache, Chiricahua and Mescalero
Use category: Food
Use sub-category: Special Food
Notes: Raw pods chewed and eaten as a delicacy.

Drug, Ear Medicine (Mescalero)

Use documented by:
Basehart, Harry W., 1974, Apache Indians XII. Mescalero Apache Subsistence Patterns and Socio-Political Organization, New York. Garland Publishing Inc., page 44

Common names: Screwbean Mesquite
Family: Fabaceae
Native American Tribe: Apache, Mescalero
Use category: Drug
Use sub-category: Ear Medicine
Notes: Pods soaked in water and used for earaches.

Drug, Ear Medicine (Western Apaches)

Use documented by:
**Buskirk, Winfred, 1986, The Western Apache: Living With the Land Before
1950, Norman. University of Oklahoma Press, page 178**

Common names: Screwbean Mesquite
Family: Fabaceae
Native American Tribe: Apache, Western
Use category: Drug
Use sub-category: Ear Medicine
Notes: Bean placed in ear for earache.

Hualapai Uses:

Fiber, Furniture

Use documented by:
**Watahomigie, Lucille J., 1982, Hualapai Ethnobotany, Peach Springs, AZ.
Hualapai Bilingual Program, Peach Springs School District #8, page 45**

Common names: Screwbean Mesquite
Family: Fabaceae
Native American Tribe: Hualapai
Use category: Fiber
Use sub-category: Furniture
Notes: Roots used to make cradleboard frames.

Food, Dried Food

Use documented by:
**Watahomigie, Lucille J., 1982, Hualapai Ethnobotany, Peach Springs, AZ.
Hualapai Bilingual Program, Peach Springs School District #8, page 45**

Common names: Screwbean Mesquite
Family: Fabaceae
Native American Tribe: Hualapai

Use category: Food
Use sub-category: Dried Food
Notes: Pods dried and stored for later use.

Isleta-Pueblo Uses:

Food-Starches

Use documented by:
Jones, Volney H., 1931, The Ethnobotany of the Isleta Indians, University of New Mexico, M.A. Thesis, page 43

Common names: Screwbean Mesquite
Family: Fabaceae
Native American Tribe: Isleta
Use category: Food
Use sub-category: Starchy Food
Notes: Pods chewed for the starch content and agreeable taste.

Kamia (Kumeyaay) Uses:

Food, Unspecified

Use documented by:
Gifford, E. W., 1931, The Kamia of Imperial Valley, Washington, D.C., U.S. Government Printing Office, page 23

Common names: Screwbean Mesquite
Family: Fabaceae
Native American Tribe: Kamia
Use category: Food
Use sub-category: Unspecified
Notes: Coiled pod used for food.

Food, Beverage

Stewart, Kenneth M., 1965, Mohave Indian Gathering of Wild Plants, Kiva 31(1):46-53, page 46

Common names: Screwbean Mesquite
Family: Fabaceae
Native American Tribe: Mojave
Use category: Food
Use sub-category: Beverage
Notes: Bean pods rotted in a pit for a month, dried, ground into a flour, and used to make a drink.

Food, Vegetable

Use documented by:
Stewart, Kenneth M., 1965, Mohave Indian Gathering of Wild Plants, Kiva 31(1):46-53, page 46

Common names: Screwbean Mesquite
Family: Fabaceae
Native American Tribe: Mojave
Use category: Food
Use sub-category: Vegetable
Notes: Bean pods used for food.

Drug, Eye Medicine

Use documented by:
Train, Percy, James R. Henrichs and W. Andrew Archer, 1941, Medicinal Uses of Plants by Indian Tribes of Nevada, Washington DC., U.S. Department of Agriculture, page 123

Common names: Screwbean Mesquite
Family: Fabaceae
Native American Tribe: Paiute
Use category: Drug
Use sub-category: Eye Medicine
Notes: Infusion of gummy exudate on bark used as an eyewash.

Food, Unspecified

Use documented by:
Murphey, Edith Van Allen, 1990, Indian Uses of Native Plants, Glenwood, IL. Meyerbooks. Originally published in 1959, page 27

Common names: Screwbean Mesquite
Family: Fabaceae
Native American Tribe: Paiute
Use category: Food
Use sub-category: Unspecified
Notes: Pounded beans used for food.

Pima Uses:

Drug, Dermatological Aid

Use documented by:
Curtin, L. S. M., 1949, By the Prophet of the Earth, Santa Fe. San Vicente Foundation, page 96

Common names: Screwbean Mesquite
Family: Fabaceae
Native American Tribe: Pima
Use category: Drug
Use sub-category: Dermatological Aid
Notes: Decoction of roots used as a wash, or powdered roots applied to sores.

Drug, Dermatological Aid

**Use documented by:
Russell, Frank, 1908, The Pima Indians, SI-BAE Annual Report #26:1-390,
page 79**

Common names: Screwbean Mesquite
Family: Fabaceae
Native American Tribe: Pima
Use category: Drug
Use sub-category: Dermatological Aid
Notes: Powdered root bark or decoction used to dress wounds.

Drug, Gynecological Aid

**Use documented by:
Curtin, L. S. M., 1949, By the Prophet of the Earth, Santa Fe. San Vicente
Foundation, page 96**

Common names: Screwbean Mesquite
Family: Fabaceae
Native American Tribe: Pima
Use category: Drug
Use sub-category: Gynecological Aid
Notes: Infusion of roots taken for troubles with menses.

Fiber, Building Material

**Use documented by:
Curtin, L. S. M., 1949, By the Prophet of the Earth, Santa Fe. San Vicente
Foundation, page 96**

Common names: Screwbean Mesquite
Family: Fabaceae
Native American Tribe: Pima

Use category: Fiber
Use sub-category: Building Material
Notes: Wood used for fence posts.

Food, Beverage

**Use documented by:
Curtin, L. S. M., 1949, By the Prophet of the Earth, Santa Fe. San Vicente
Foundation, page 96**

Common names: Screwbean Mesquite
Family: Fabaceae
Native American Tribe: Pima
Use category: Food
Use sub-category: Beverage
Notes: Beans ground, mixed with water, and made into a sweet and nourishing
beverage.

Food, Beverage

**Use documented by:
Hrdlicka, Ales, 1908, Physiological and Medical Observations Among the
Indians of Southwestern United States and Northern Mexico, SI-BAE
Bulletin #34:1-427, page 261**

Common names: Screwbean Mesquite
Family: Fabaceae
Native American Tribe: Pima
Use category: Food
Use sub-category: Beverage
Notes: Beans sun-dried, pounded into meal, mixed with cold water, and used as a
drink.

Food, Candy

**Use documented by:
Curtin, L. S. M., 1949, By the Prophet of the Earth, Santa Fe. San Vicente
Foundation, page 96**

Common names: Screwbean Mesquite
Family: Fabaceae
Native American Tribe: Pima
Use category: Food
Use sub-category: Candy
Notes: Fresh, sugary pods chewed by children.

Food, Forage

**Use documented by:
Curtin, L. S. M., 1949, By the Prophet of the Earth, Santa Fe. San Vicente
Foundation, page 96**

Common names: Screwbean Mesquite
Family: Fabaceae
Native American Tribe: Pima
Use category: Food
Use sub-category: Forage
Notes: Pods and foliage eaten by grazing animals.

Food, Staple

**Use documented by:
Russell, Frank, 1908, The Pima Indians, SI-BAE Annual Report #26:1-390,
page 75**

Common names: Screwbean Mesquite
Family: Fabaceae
Native American Tribe: Pima
Use category: Food

Use sub-category: Staple
Notes: Beans pit cooked, dried, pounded, and eaten as pinole.

Other, Fuel

**Use documented by:
Curtin, L. S. M., 1949, By the Prophet of the Earth, Santa Fe. San Vicente
Foundation, page 96**

Common names: Screwbean Mesquite
Family: Fabaceae
Native American Tribe: Pima
Use category: Other
Use sub-category: Fuel
Notes: Wood used for fuel

Pima (Gila River) Uses:

Food, Snack Food

**Use documented by:
Rea, Amadeo M., 1991, Gila River Pima Dietary Reconstruction, Arid
Lands Newsletter 31:3- 10, page 5**

Common names: Screwbean Mesquite
Family: Fabaceae
Native American Tribe: Pima, Gila River
Use category: Food
Use sub-category: Snack Food
Notes: Catkins eaten as a snack food by all age groups.

Food, Snack Food

**Use documented by:
Rea, Amadeo M., 1991, Gila River Pima Dietary Reconstruction, Arid
Lands Newsletter 31:3- 10, page 5**

Common names: Screwbean Mesquite
Family: Fabaceae
Native American Tribe: Pima, Gila River
Use category: Food
Use sub-category: Snack Food
Notes: Sap eaten as a snack food by all age groups.

Food, Staple

**Use documented by:
Rea, Amadeo M., 1991, Gila River Pima Dietary Reconstruction, Arid
Lands Newsletter 31:3- 10, page 5**

Common names: Screwbean Mesquite
Family: Fabaceae
Native American Tribe: Pima, Gila River
Use category: Food
Use sub-category: Staple
Notes: Beans used to make flour.

Food, Staple

**Use documented by:
Rea, Amadeo M., 1991, Gila River Pima Dietary Reconstruction, Arid
Lands Newsletter 31:3- 10, page 7**

Common names: Screwbean Mesquite
Family: Fabaceae
Native American Tribe: Pima, Gila River
Use category: Food

Use sub-category: Staple
Notes: Fruit used as a staple food.

Other, Season Indicator

**Use documented by:
Rea, Amadeo M., 1991, Gila River Pima Dietary Reconstruction, Arid Lands Newsletter 31:3- 10, page 6**

Common names: Screwbean Mesquite
Family: Fabaceae
Native American Tribe: Pima, Gila River
Use category: Other
Use sub-category: Season Indicator
Notes: Leaves used as a sign that planted crops would be safe from freezing weather.

Tewa-Pueblo Uses:

Drug, Ear Medicine

**Use documented by:
Robbins, W.W., J.P. Harrington, and B. Freire-Marreco, 1916, Ethnobotany of the Tewa Indians, SI-BAE Bulletin #55, page 69**

Common names: Screwbean Mesquite
Family: Fabaceae
Native American Tribe: Tewa
Use category: Drug
Use sub-category: Ear Medicine
Notes: Pods twisted into the ear for an earache.

Cirsium neomexicanum

New Mexico Thistle

Drug, Febrifuge

Use documented by:
Elmore, Francis H., 1944, Ethnobotany of the Navajo, Santa Fe, NM. School of American Research, page 96

Common names: New Mexico Thistle
Family: Asteraceae
Native American Tribe: Navajo
Use category: Drug
Use sub-category: Febrifuge
Notes: Plant used for chills and fevers.

Drug, Eye Medicine

Use documented by:
Vestal, Paul A., 1952, The Ethnobotany of the Ramah Navaho, Papers of the Peabody Museum of American Archaeology and Ethnology 40(4):1-94, page 50

Common names: New Mexico Thistle
Family: Asteraceae
Native American Tribe: Navajo, Ramah
Use category: Drug
Use sub-category: Eye Medicine
Notes: Cold infusion of root used as a wash for eye diseases.

Drug, Panacea

Use documented by:
Vestal, Paul A., 1952, The Ethnobotany of the Ramah Navaho, Papers of the Peabody Museum of American Archaeology and Ethnology 40(4):1-94, page 50

Common names: New Mexico Thistle
Family: Asteraceae
Native American Tribe: Navajo, Ramah
Use category: Drug
Use sub-category: Panacea
Notes: Cold infusion of plant taken when one 'feels bad all over.'

Drug, Panacea

Use documented by:
Vestal, Paul A., 1952, The Ethnobotany of the Ramah Navaho, Papers of the Peabody Museum of American Archaeology and Ethnology 40(4):1-94, page 50

Common names: New Mexico Thistle
Family: Asteraceae
Native American Tribe: Navajo, Ramah
Use category: Drug
Use sub-category: Panacea
Notes: Root used as a 'life medicine.'

Drug, Veterinary Aid

Use documented by:
Vestal, Paul A., 1952, The Ethnobotany of the Ramah Navaho, Papers of the Peabody Museum of American Archaeology and Ethnology 40(4):1-94, page 50

Common names: New Mexico Thistle
Family: Asteraceae
Native American Tribe: Navajo, Ramah
Use category: Drug
Use sub-category: Veterinary Aid
Notes: Cold infusion of root used as a wash for livestock with eye diseases.

Food, Unspecified

Use documented by:
Gifford, E.W., 1936, Northeastern and Western Yavapai,
University of California Publications in American Archaeology and
Ethnology 34:247-345, page 256

Common names: New Mexico Thistle
Family: Asteraceae
Native American Tribe: Yavapai
Use category: Food
Use sub-category: Unspecified
Notes: Raw, peeled stems used for food.

Forestiera pubescens

New Mexico Olive/ Desert Olive/ Stretchberry

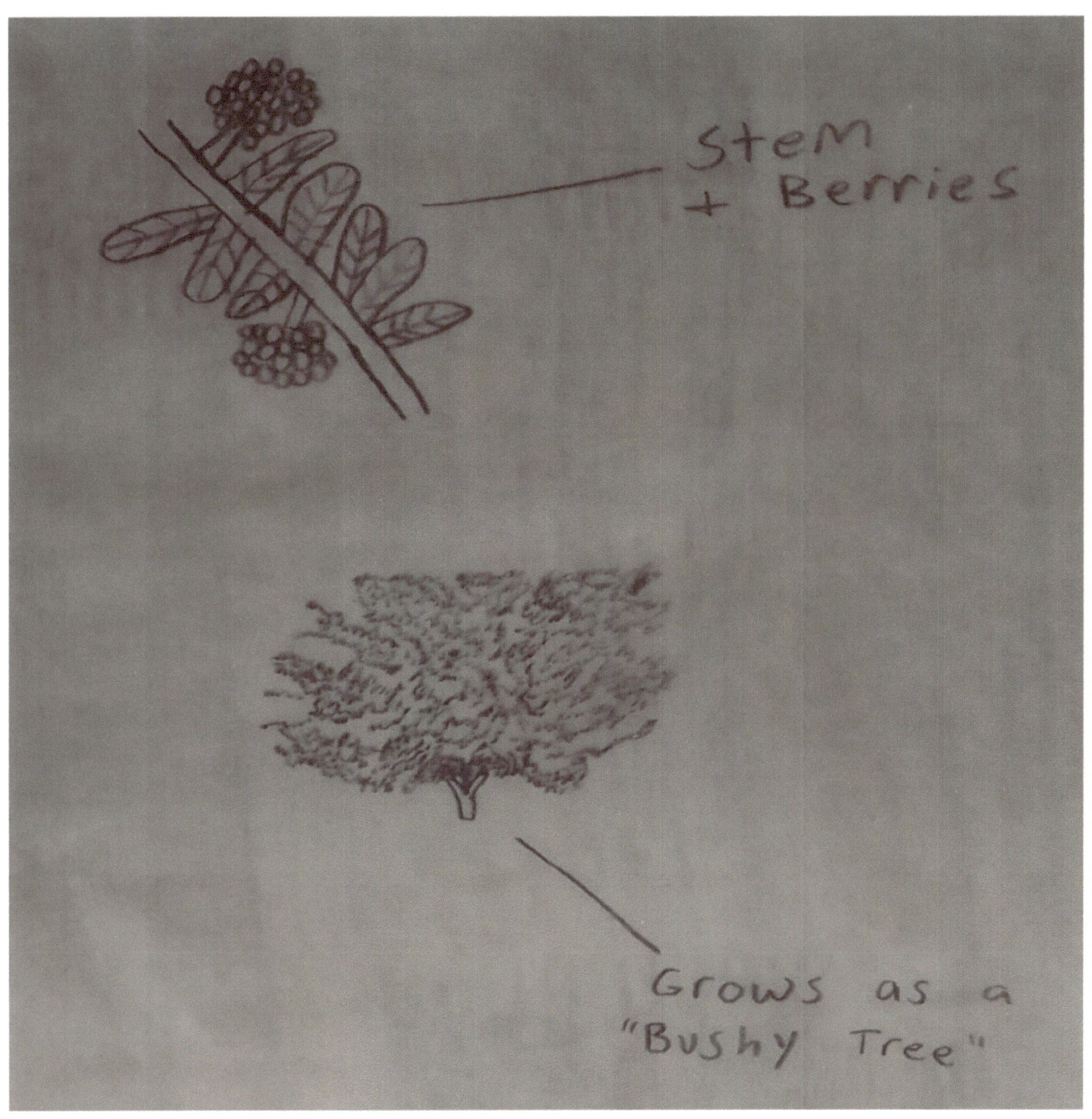

Apache (Chiricahua/Mescalero) Uses:

Food, Fruit

Use documented by:
Castetter, Edward F. and M. E. Opler, 1936, Ethnobiological Studies in the American Southwest III. The Ethnobiology of the Chiricahua and Mescalero Apache, University of New Mexico Bulletin 4(5):1-63, page 44

Common names: New Mexico Olive, Desert Olive, Stretchberry
Family: Oleaceae
Native American Tribe: Apache, Chiricahua and Mescalero
Use category: Food
Use sub-category: Fruit
Notes: Raw fruits occasionally eaten as food.

Hopi Uses:

Other, Ceremonial Items

Use documented by:
Colton, Harold S., 1974, Hopi History and Ethnobotany, IN D. A. Horr (ed.) Hopi Indians. Garland: New York, page 319

Common names: New Mexico Olive, Desert Olive, Stretchberry
Family: Oleaceae
Native American Tribe: Hopi
Use category: Other
Use sub-category: Ceremonial Items
Notes: Used to make pahos (prayer sticks).

Other, Tools

Use documented by:
Colton, Harold S., 1974, Hopi History and Ethnobotany, IN D. A. Horr (ed.) Hopi Indians. Garland: New York, page 319

Common names: New Mexico Olive, Desert Olive, Stretchberry
Family: Oleaceae
Native American Tribe: Hopi
Use category: Other
Use sub-category: Tools
Notes: Used for digging stick.

Isleta-Pueblo Uses:

Other, Water Indicator

Use documented by:
Jones, Volney H., 1931, The Ethnobotany of the Isleta Indians, University of New Mexico, M.A. Thesis, page 29

Common names: New Mexico Olive, Desert Olive, Stretchberry
Family: Oleaceae
Native American Tribe: Isleta
Use category: Other
Use sub-category: Water Indicator
Notes: Large shrubs considered water indicators because wells dug where plants grew always produced water.

Jemez-Pueblo Uses:

Other, Ceremonial Items

Use documented by:
Cook, Sarah Louise, 1930, The Ethnobotany of Jemez Indians., University of New Mexico, M.A. Thesis, page 22

Common names: New Mexico Olive, Desert Olive, Stretchberry
Family: Oleaceae
Native American Tribe: Jemez
Use category: Other

Use sub-category: Ceremonial Items
Notes: Berry juice mixed with white clay and used as purple body paint for summer dances.

Navajo Uses:

Other, Ceremonial Items

Use documented by:
Elmore, Francis H., 1944, Ethnobotany of the Navajo, Santa Fe, NM. School of American Research, page 68

Common names: New Mexico Olive, Desert Olive, Stretchberry
Family: Oleaceae
Native American Tribe: Navajo
Use category: Other
Use sub-category: Ceremonial Items
Notes: Used to make prayer sticks.

Navajo (Ramah) Uses:

Drug, Ceremonial Medicine

Use documented by:
Vestal, Paul A., 1952, The Ethnobotany of the Ramah Navaho, Papers of the Peabody Museum of American Archaeology and Ethnology 40(4):1-94, page 39

Common names: New Mexico Olive, Desert Olive, Stretchberry
Family: Oleaceae
Native American Tribe: Navajo, Ramah
Use category: Drug
Use sub-category: Ceremonial Medicine
Notes: Leaves used as a ceremonial emetic.

Drug, Disinfectant

Use documented by:
Vestal, Paul A., 1952, The Ethnobotany of the Ramah Navaho, Papers of the Peabody Museum of American Archaeology and Ethnology 40(4):1-94, page 39

Common names: New Mexico Olive, Desert Olive, Stretchberry
Family: Oleaceae
Native American Tribe: Navajo, Ramah
Use category: Drug
Use sub-category: Disinfectant
Notes: Plant used for 'bear infection.'

Drug, Emetic

Use documented by:
Vestal, Paul A., 1952, The Ethnobotany of the Ramah Navaho, Papers of the Peabody Museum of American Archaeology and Ethnology 40(4):1-94, page 39

Common names: New Mexico Olive, Desert Olive, Stretchberry
Family: Oleaceae
Native American Tribe: Navajo, Ramah
Use category: Drug
Use sub-category: Emetic
Notes: Leaves used as a ceremonial emetic.

Other, Ceremonial Items

Use documented by:
Vestal, Paul A., 1952, The Ethnobotany of the Ramah Navaho, Papers of the Peabody Museum of American Archaeology and Ethnology 40(4):1-94, page 39

Common names: New Mexico Olive, Desert Olive, Stretchberry
Family: Oleaceae
Native American Tribe: Navajo, Ramah
Use category: Other
Use sub-category: Ceremonial Items
Notes: Stem used to make Evilway big hoop.

Psorothamnus scoparius

Broom Dalea

Drug, Dermatological Aid

Use documented by:
**Swank, George R., 1932, The Ethnobotany of the Acoma and Laguna
Indians, University of New Mexico, M.A. Thesis, page 57**

Common names: Broom Dalea
Family: Fabaceae
Native American Tribe: Keres, Western
Use category: Drug
Use sub-category: Dermatological Aid
Notes: Infusion of plant rubbed on spider bites.

Drug, Emetic

Use documented by:
**Swank, George R., 1932, The Ethnobotany of the Acoma and Laguna
Indians, University of New Mexico, M.A. Thesis, page 57**

Common names: Broom Dalea
Family: Fabaceae
Native American Tribe: Keres, Western
Use category: Drug
Use sub-category: Emetic
Notes: Infusion of plant used as an emetic.

Drug, Gastrointestinal Aid

Use documented by:
**Swank, George R., 1932, The Ethnobotany of the Acoma and Laguna
Indians, University of New Mexico, M.A. Thesis, page 57**

Common names: Broom Dalea
Family: Fabaceae

Native American Tribe: Keres, Western
Use category: Drug
Use sub-category: Gastrointestinal Aid
Notes: Infusion of plant used for stomach trouble.

Vitis arizonica
Arizona Grape/Canyon Grape

Apache (Chiricahua/Mescalero) Uses:

Food, Dried Food

Use documented by:
Castetter, Edward F. and M. E. Opler, 1936, Ethnobiological Studies in the American Southwest III. The Ethnobiology of the Chiricahua and Mescalero Apache, University of New Mexico Bulletin 4(5):1-63, page 44

Common names: Arizona Grape, Canyon Grape
Family: Vitaceae
Native American Tribe: Apache, Chiricahua and Mescalero
Use category: Food
Use sub-category: Dried Food
Notes: Fruit dried and eaten like raisins.

Food, Fruit

Use documented by:
Basehart, Harry W., 1974, Apache Indians XII. Mescalero Apache Subsistence Patterns and Socio-Political Organization, New York. Garland Publishing Inc., page 50

Common names: Arizona Grape, Canyon Grape
Family: Vitaceae
Native American Tribe: Apache, Mescalero
Use category: Food
Use sub-category: Fruit
Notes: Fruits eaten fresh.

Food, Beverage

Use documented by:
Buskirk, Winfred, 1986, The Western Apache: Living With the Land Before
1950, Norman. University of Oklahoma Press, page 190

Common names: Arizona Grape, Canyon Grape
Family: Vitaceae
Native American Tribe: Apache, Western
Use category: Food
Use sub-category: Beverage
Notes: Juice boiled to make wine.

Food, Dried Food

Use documented by:
Buskirk, Winfred, 1986, The Western Apache: Living With the Land Before
1950, Norman. University of Oklahoma Press, page 190

Common names: Arizona Grape, Canyon Grape
Family: Vitaceae
Native American Tribe: Apache, Western
Use category: Food
Use sub-category: Dried Food
Notes: Berries pounded, dried, and stored in sacks.

Food, Fruit

Use documented by:
Buskirk, Winfred, 1986, The Western Apache: Living With the Land Before
1950, Norman. University of Oklahoma Press, page 190

Common names: Arizona Grape, Canyon Grape
Family: Vitaceae

Native American Tribe: Apache, Western
Use category: Food
Use sub-category: Fruit
Notes: Ripe berries eaten raw.

Havasupai Uses:

Food, Fruit

Use documented by:
Weber, Steven A. and P. David Seaman, 1985, Havasupai Habitat: A. F.
Whiting's Ethnography of a Traditional Indian Culture, Tucson. The
University of Arizona Press, page 231

Common names: Arizona Grape, Canyon Grape
Family: Vitaceae
Native American Tribe: Havasupai
Use category: Food
Use sub-category: Fruit
Notes: Fruit used for food.

Other, Toys and Games

Use documented by:
Weber, Steven A. and P. David Seaman, 1985, Havasupai Habitat: A. F.
Whiting's Ethnography of a Traditional Indian Culture, Tucson. The
University of Arizona Press, page 231

Common names: Arizona Grape, Canyon Grape
Family: Vitaceae
Native American Tribe: Havasupai
Use category: Other
Use sub-category: Toys and Games
Notes: Vines used to make the hoop of the hoop and pole game.

Isleta-Pueblo Uses:

Food, Fruit

Use documented by:
Jones, Volney H., 1931, The Ethnobotany of the Isleta Indians, University of New Mexico, M.A. Thesis, page 44

Common names: Arizona Grape, Canyon Grape
Family: Vitaceae
Native American Tribe: Isleta
Use category: Food
Use sub-category: Fruit
Notes: Fruit considered an important part of the diet.

Jemez-Pueblo Uses:

Food, Fruit

Use documented by:
Cook, Sarah Louise, 1930, The Ethnobotany of Jemez Indians, University of New Mexico, M.A. Thesis, page 28

Common names: Arizona Grape, Canyon Grape
Family: Vitaceae
Native American Tribe: Jemez
Use category: Food
Use sub-category: Fruit
Notes: Grapes used for food.

Other, Ceremonial Items

Use documented by:
Cook, Sarah Louise, 1930, The Ethnobotany of Jemez Indians, University of New Mexico, M.A. Thesis, page 28

Common names: Arizona Grape, Canyon Grape
Family: Vitaceae
Native American Tribe: Jemez
Use category: Other
Use sub-category: Ceremonial Items
Notes: Berry juice mixed with white clay and used as a body paint for dancers.

Navajo Uses:

Drug, Love Medicine

Use documented by:
Elmore, Francis H., 1944, Ethnobotany of the Navajo, Santa Fe, NM. School of American Research, page 62

Common names: Arizona Grape, Canyon Grape
Family: Vitaceae
Native American Tribe: Navajo
Use category: Drug
Use sub-category: Love Medicine
Notes: Vine used to make a cross, and put on top of the basket of cornmeal and paper bread offered in courtship.

Panicum capillare

Witchgrass

Food, Bread and Cake

Use documented by:
Vestal, Paul A, 1940, Notes on a Collection of Plants from the Hopi Indian Region of Arizona Made by J. G. Owens in 1891, Botanical Museum Leaflets (Harvard University) 8(8):153-168, page 159

Common names: Witchgrass
Family: Poaceae
Native American Tribe: Hopi
Use category: Food
Use sub-category: Bread and Cake
Notes: Ground seed meal used to make bread.

Food, Staple

Use documented by:
Fewkes, J. Walter, 1896, A Contribution to Ethnobotany, American Anthropologist 9:14-21, page 17

Common names: Witchgrass
Family: Poaceae
Native American Tribe: Hopi
Use category: Food
Use sub-category: Staple
Notes: Seeds ground and mixed with corn meal.

Drug, Emetic

Use documented by:
Swank, George R., 1932, The Ethnobotany of the Acoma and Laguna Indians, University of New Mexico, M.A. Thesis, page 57

Common names: Witchgrass
Family: Poaceae
Native American Tribe: Keres, Western
Use category: Drug
Use sub-category: Emetic
Notes: Infusion of leaves used as an emetic before breakfast.

Navajo Uses:

Food, Unspecified

Use documented by:
Elmore, Francis H., 1944, Ethnobotany of the Navajo, Santa Fe, NM. School of American Research, page 26

Common names: Witchgrass
Family: Poaceae
Native American Tribe: Navajo
Use category: Food
Use sub-category: Unspecified
Notes: Seeds used for food.

Navajo (Ramah) Uses:

Food, Fodder

Use documented by:
Vestal, Paul A., 1952, The Ethnobotany of the Ramah Navaho, Papers of the Peabody Museum of American Archaeology and Ethnology 40(4):1-94, page 17

Common names: Witchgrass
Family: Poaceae
Native American Tribe: Navajo, Ramah

Use category: Food
Use sub-category: Fodder
Notes: Used for sheep and horse feed.

Tewa (Pueblo) Uses:

Fiber, Brushes and Brooms

Use documented by:
Robbins, W.W., J.P. Harrington, and B. Freire-Marreco, 1916, Ethnobotany of the Tewa Indians, SI-BAE Bulletin #55, page 64

Common names: Witchgrass
Family: Poaceae
Native American Tribe: Tewa
Use category: Fiber
Use sub-category: Brushes and Brooms
Notes: Grass made into brooms and used to clean metates and metate boxes.

Phragmites australis

Common Reed

Drug, Antidiarrheal

Use documented by:
Reagan, Albert B., 1929, Plants Used by the White Mountain Apache
Indians of Arizona, Wisconsin Archeologist 8:143-61, page 159

Common names: Common Reed
Family: Poaceae
Native American Tribe: Apache, White Mountain
Use category: Drug
Use sub-category: Antidiarrheal
Notes: Root used for diarrhea and kindred diseases.

Other, Hunting and Fishing Item

Use documented by:
Reagan, Albert B., 1929, Plants Used by the White Mountain Apache
Indians of Arizona, Wisconsin Archeologist 8:143-61, page 159

Common names: Common Reed
Family: Poaceae
Native American Tribe: Apache, White Mountain
Use category: Other
Use sub-category: Hunting and Fishing Item
Notes: Reeds used as arrow shafts for hunting small birds with arrows.

Other, Smoke Plant

Use documented by:
Reagan, Albert B., 1929, Plants Used by the White Mountain Apache
Indians of Arizona, Wisconsin Archeologist 8:143-61, page 159

Common names: Common Reed
Family: Poaceae

Native American Tribe: Apache, White Mountain
Use category: Other
Use sub-category: Smoke Plant
Notes: Reeds filled with tobacco and used as cigarettes.

Other, Smoking Tools

Use documented by:
**Reagan, Albert B., 1929, Plants Used by the White Mountain Apache
Indians of Arizona, Wisconsin Archeologist 8:143-61, page 159**

Common names: Common Reed
Family: Poaceae
Native American Tribe: Apache, White Mountain
Use category: Other
Use sub-category: Smoking Tools
Notes: Reeds used to make pipe stems.

Cocopah Uses:

Other, Smoking Tools

Use documented by:
**Castetter, Edward F. and Willis H. Bell, 1951, Yuman Indian Agriculture,
Albuquerque: University of New Mexico Press, page 122**

Common names: Common Reed
Family: Poaceae
Native American Tribe: Cocopah
Use category: Other
Use sub-category: Smoking Tools
Notes: Tubular internodes used to smoke tobacco.

Havasupai Uses:

Fiber, Mats, Rugs, and Bedding

Use documented by:
Weber, Steven A. and P. David Seaman, 1985, Havasupai Habitat: A.F.
Whiting's Ethnography of a Traditional Indian Culture, Tucson. The
University of Arizona Press, page 209

Common names: Common Reed
Family: Poaceae
Native American Tribe: Havasupai
Use category: Fiber
Use sub-category: Mats, Rugs, and Bedding
Notes: Stems used to make mats for drying yucca fruit pulp, baked mescal,
peaches, or figs.

Data Analysis of Documented Botanical Uses

Below, we will review the aforementioned botanical case study data from both a
qualitative and quantitative perspective, while being brief and to the point. In this
way, all audiences can easily understand the results. I hope that in the near future,
this information will be used to raise awareness of Native American culture and
promote conservation efforts of such botanical species.

Total Botanical Species (Baseline Data): 12

Uses For Botanical Species	% Of Human Resource Use For Total Botanical Species (by category)	Botanical Species Present For Human Resource Use (by category)	Tribes Utilizing Botanical Resources (human resource use by category)
Drug (ethnomedicine)	10/12 =83.33%	1.Silverleaf Nightshade 2.Prairie Sunflower 3.Yerba Mansa 4.Screwbean Mesquite 5.New Mexico Thistle 6.New Mexico Olive/Desert Olive/ Stretchberry 7.Broom Dalea 8.Arizona Grape/ Canyon Grape 9.Witchgrass 10.Common Reed	1.Navajo 2.Zuni 3.Cochiti-Pueblo 4.Isleta-Pueblo 5.Keresan-Pueblo 6.Hopi 7.Pima 8.Apache-White Mountain 9.Other Groups: Spanish American 10.Havasupai 11.Kamia-Kumeyaay 12.Keres-Pueblo (Western) 13.Paiute 14.Papago 15.Pima-Gila River 16.Apache-Chiricahua & Mescalero 17.Hualapai 18.Mojave 19.Tewa-Pueblo 20.Navajo-Ramah 21.Yavapai 22.Jemez-Pueblo 23.Apache Western (Western Apache's) 24.Cocopah

Total Botanical Species (Baseline Data): 12

Uses For Botanical Species	% Of Human Resource Use For Total Botanical Species (by category)	Botanical Species Present For Human Resource Use (by category)	Tribes Utilizing Botanical Resources (human resource use by category)
Food (subsistence)	10/12=83.33%	1.Silverleaf Nightshade 2.Rio Grande Cottonwood 3.Prairie Sunflower 4.Mexican Devil-Weed 5.Yerba Mansa 6.Screwbean Mesquite 7.New Mexico Thistle 8.New Mexico Olive/Desert Olive/ Stretchberry 9.Arizona Grape/ Canyon Grape 10.Witchgrass	1.Navajo 2.Zuni 3.Cochiti-Pueblo 4.Isleta-Pueblo 5.Keresan Pueblo 6.Hopi 7.Pima 8.Apache-White Mountain 9.Other Groups: Spanish American 10.Acoma-Pueblo 11.Apache-Chiricuahua & Mescalero 12.Jemez Pueblo 13.Keres-Pueblo (Western) 14.Laguna Pueblo 15.Tewa-Pueblo 16.Havasupai 17.Mojave 18.Kamia-Kumeyaay 19.Paiute 20.Papago 21.Hualapai 22.Pima-Gila River 23.Navajo-Ramah 24.Yavapai 25.Apache-Western (Western Apache's)

Total Botanical Species (Baseline Data): 12

Uses For Botanical Species	% Of Human Resource Use For Total Botanical Species (by category)	Botanical Species Present For Human Resource Use (by category)	Tribes Utilizing Botanical Resources (human resource use by category)
Jewelry (adornment)	1/12=8.33%	1.Silverleaf Nightshade	1.Navajo 2.Zuni 3.Cochiti-Pueblo 4.Isleta-Pueblo 5.Keresan-Pueblo 6.Hopi 7.Pima 8.Apache-White Mountain 9.Other Groups:Spanish American
Fiber (constructing of cultural objects)	4/12=33.33%	1.Rio Grande Cottonwood 2.Screwbean Mesquite 3.Witchgrass 4.Common Reed	1.Acoma-Pueblo 2.Apache-Chiricuahua & Mescalero 3.Apache-White Mountain 4.Isleta-Pueblo 5.Jemez-Pueblo 6.Keres-Pueblo (Western) 7.Laguna-Pueblo 8.Navajo 9.Pima 10.Tewa-Pueblo 11.Zuni 12.Hualapai 13.Kamia-Kumeyaay 14.Mojave 15.Paiute 16.Pima-Gila River 17.Hopi 18.Navajo-Ramah 19.Cocopah 20.Havasupai

Total Botanical Species (Baseline Data): 12

Uses For Botanical Species	% Of Human Resource Use For Total Botanical Species (by category)	Botanical Species Present For Human Resource Use (by category)	Tribes Utilizing Botanical Resources (human resource use by category)
Cash Crop (means of financially sustaining one's self & family)	1/12=8.33%	1.Rio Grande Cottonwood	1.Acoma-Pueblo 2.Apache-Chiricahua & Mescalero 3.Apache-White Mountain 4.Isleta-Pueblo 5.Jemez-Pueblo 6.Keres-Pueblo (Western) 7.Laguna-Pueblo 8.Navajo 9.Pima 10.Tewa-Pueblo 11.Zuni
Ceremonial Items (used for ceremony)	4/12=33.33%	1.Rio Grande Cottonwood 2.Prairie Sunflower 3.New Mexico Olive/Desert Olive/ Stretchberry 4.Arizona Grape/ Canyon Grape	1.Acoma-Pueblo 2.Apache-Chiricahua & Mescalero 3.Apache-White Mountain 4.Isleta-Pueblo 5.Jemez-Pueblo 6.Keres-Pueblo (Western) 7.Laguna-Pueblo 8.Navajo 9.Pima 10.Tewa-Pueblo 11. Zuni 12.Havasupai 13.Hopi 14.Navajo-Ramah 15.Apache-Western (Western Apache's)

Total Botanical Species (Baseline Data): 12

Uses For Botanical Species	% Of Human Resource Use For Total Botanical Species (by category)	Botanical Species Present For Human Resource Use (by category)	Tribes Utilizing Botanical Resources (human resource use by category)
Fuel (heating)	2/12=16.66%	1.Rio Grande Cottonwood 2.Screwbean Mesquite	1.Acoma-Pueblo 2.Apache-Chiricahua & Mescalero 3.Apache-White Mountain 4.Isleta-Pueblo 5.Jemez-Pueblo 6.Keres-Pueblo (Western) 7.Laguna-Pueblo 8.Navajo 9.Pima 10.Tewa-Pueblo 11.Zuni 12.Hualapai 13.Kamia-Kumeyaay 14.Mojave 15.Paiute 16.Pima 17.Pima-Gila River
Containers (botanical species used to create holding containers for storage purposes)	1/12=8.33%	1.Rio Grande Cottonwood	1.Acoma-Pueblo 2.Apache-Chiricahua & Mescalero 3.Apache-White Mountain 4.Isleta-Pueblo 5.Jemez-Pueblo 6.Keres (Western)-Pueblo 7.Laguna-Pueblo 8.Navajo 9.Pima 10.Tewa-Pueblo 11.Zuni

Total Botanical Species (Baseline Data): 12

Uses For Botanical Species	% Of Human Resource Use For Total Botanical Species (by category)	Botanical Species Present For Human Resource Use (by category)	Tribes Utilizing Botanical Resources (human resource use by category)
Tools	2/12=16.66%	1.Rio Grande Cottonwood 2.New Mexico Olive/Desert Olive/ Stretchberry	1.Acoma-Pueblo 2.Apache-Chiricahua & Mescalero 3.Apache-White Mountain 4.Isleta-Pueblo 5.Jemez-Pueblo 6.Keres (Western)-Pueblo 7.Laguna-Pueblo 8.Navajo 9.Pima 10.Tewa-Pueblo 11.Zuni 12.Hopi 13.Navajo-Ramah
Decorations (beautifying surroundings)	2/12=16.66%	1.Rio Grande Cottonwood 2.Prairie Sunflower	1.Acoma-Pueblo 2.Apache-Chiricahua & Mescalero 3.Apache-White Mountain 4.Isleta-Pueblo 5.Jemez-Pueblo 6.Keres (Western)-Pueblo 7.Laguna-Pueblo 8.Navajo 9.Pima 10.Tewa-Pueblo 11.Zuni 12.Havasupai 13.Hopi

Total Botanical Species (Baseline Data): 12

Uses For Botanical Species	% Of Human Resource Use For Total Botanical Species (by category)	Botanical Species Present For Human Resource Use (by category)	Tribes Utilizing Botanical Resources (human resource use by category)
Season Indicator (seasonal indicators in regards to weather patterns & agriculture)	2/12=16.66%	1.Prairie Sunflower 2.Screwbean Mesquite	1.Havasupai 2.Hopi 3.Navajo 4.Apache-Chiricahua & Mescalero 5.Hualapai 6.Isleta-Pueblo 7.Kamia-Kumeyaay 8.Mojave 9.Paiute 10.Pima 11.Pima-Gila River 12.Tewa-Pueblo
Water Indicator (indicating good localized sources of water present)	1/12=8.33%	1.New Mexico Olive/Desert Olive/ Stretchberry	1.Apache-Chiricahua & Mescalero 2.Hopi 3.Isleta-Pueblo 4.Jemez Pueblo 5.Navajo 6.Navajo-Ramah
Toys & Games	1/12=8.33%	1. Arizona Grape/ Canyon Grape	1.Apache-Chiricahua & Mescalero 2.Apache-Western (Western Apache's) 3.Havasupai 4.Isleta-Pueblo 5.Jemez-Pueblo 6.Navajo
Hunting & Fishing Item	1/12=8.33%	1.Common Reed	1.Apache-White Mountain 2.Cocopah 3.Havasupai

Total Botanical Species (Baseline Data): 12

Uses For Botanical Species	% Of Human Resource Use For Total Botanical Species (by category)	Botanical Species Present For Human Resource Use (by category)	Tribes Utilizing Botanical Resources (human resource use by category)
Smoke Plant (plants used for smoking consumption)	1/12=8.33%	1.Common Reed	1.Apache-White Mountain 2.Cocopah 3.Havasupai
Smoking Tools	1/12=8.33%	1.Common Reed	1.Apache-White Mountain 2.Cocopah 3.Havasupai

Plant Species Utilized In Study:

1. Silverleaf Nightshade
2. Rio Grande Cottonwood
3. Prairie Sunflower
4. Mexican Devil-Weed
5. Yerba Mansa
6. Screwbean Mesquite
7. New Mexico Thistle
8. New Mexico Olive/Desert Olive/Stretchberry
9. Broom Dalea
10. Arizona Grape/Canyon Grape
11. Witchgrass
12. Common Reed

Tribes Documented For Study:

1. Navajo
2. Zuni
3. Cochiti-Pueblo
4. Isleta-Pueblo
5. Keresan-Pueblo
6. Hopi
7. Pima
8. Apache-White Mountain
9. Acoma-Pueblo
10. Apache-Chiricahua & Mescalero
11. Jemez-Pueblo
12. Keres-Pueblo (Western)
13. Laguna-Pueblo
14. Tewa-Pueblo
15. Havasupai
16. Mojave
17. Kamia-Kumeyaay
18. Paiute
19. Papago
20. Hualapai
21. Pima-Gila River
22. Navajo-Ramah
23. Yavapai
24. Apache-Western (Western Apache's)
25. Cocopah
26. Other Groups: Spanish American

Integrating Bio-Cultural Diversity into Wildlife and Natural Heritage Conservation (Ecosystem Services)

In this section, I will discuss how to "integrate" potential ways of measuring and establishing "Bio-Cultural Diversity" through the field of Ecosystem Services. We can measure and estimate how various plants and animals could directly impact cultural diversity through known usages for these plants and animals (Ecosystem Services). For the purposes of this section, I will use my own mathematical expressions/equations to show direct correlations and interactions between plants, animals, and human culture, as well as showing what this could ultimately mean for both natural resources and cultural heritage. For this section, I will give "theoretical examples" of how this could be integrated.

"Biocultural Diversity" is any type of biological diversity that also plays a direct role and link between the natural world and human culture. Various ways in which biological/natural resources can interact with human culture includes (but is not limited to) linguistics, tangible heritage culture, and intangible heritage culture. Think about the various Native American tribes in this country. What do you think would happen to their cultural heritages if the plants or animals they traditionally used to practice their cultures became endangered or went extinct? Not only would their cultures become endangered, but they would also experience linguistic loss (especially if species went extinct). If tribes use specific words to describe particular plants and animals that could potentially go extinct, so too, would their reason to use those words to describe those natural resources disappear, and thereby they would lose parts of their languages and themselves.

"Ecosystem Services" are any type of service that both plants and animal species can provide to us for daily living and pleasure. Examples include (but are not limited to) spirituality, food, clothing, housing, recreation, hunting, fishing, diving, boating, etc. Ecosystem Services are an essential part of nature, and are also an essential part of everyday human health and well-being. Let's look at some examples below of how to "integrate" potential ways of measuring and establishing "Bio-Cultural Diversity" through the field of Ecosystem Services.

First, let's list the various types of "Bio-Cultural Diversity" that can be estimated and measured through "Ecosystem Services."

Various "Ecosystem Services" that can be measured, estimated, and quantified for "Bio-Cultural Diversity" Loss.

1. Clothing
2. Shelter
3. Food
4. Spirituality
5. Transportation
6. Recreation

Bridgewater, P. and Upadhaya, S. (2022, September 7). *What is Biocultural Diversity, and why does it Matter?* The Conversation. Retrieved September 24, 2022, from https://theconversation.com/what-is-biocultural-diversity-and-why-does-it-matter-168881

Seele, B. C., K. J. Esler, and A. B. Cunningham. 2019. Biocultural Diversity: A Mongolian Case Study. *Ecology and Society* 24(4):27. https://doi.org/10.5751/ES-11207-240427

Let's assume that all six of these Ecosystem Services categories are equal to and add up to 100% of human cultural diversity. Therefore, each category makes up 16.666666666666667 % of all human cultural diversity through the use of natural resources/wildlife resources by various cultures around the world.

ES = Ecosystem Services

@ = At

% = Percent

- = Minus

= -- Equals

PS #1 = Plant Species #1

BCD HC #1=Bio-Cultural Diversity for Human Culture #1

Example #1:

Let's say that plant species #1 is being used for clothing, food, shelter, and spirituality for human culture #1. Plant Species #1 is being used for over 66.6666666666668 % of human culture for human culture #1. Let's also say that plant species #1 has just gone extinct. This now means that due to plant species #1 going extinct, human culture #1 has just experienced a Bio-Cultural Diversity loss of over 66.6666666666668 %, or a loss of cultural heritage by over 2/3 for the entire human culture being studied. We could also write this out as the following mathematical expression/equation:

ES @ 100 % - PS #1 (16.6666666666667 %) (4) = BCD HC #1 Remaining

ES @ 100 % - PS #1 (66.6666666666668 %) = 33.3333333333332 %.

This could also be written out as:

Ecosystem Services @ 100 % - Plant Species #1 (16.6666666666667 %) (4) = Bio-Cultural Diversity for Human Culture #1 Remaining

Ecosystem Services @ 100 % - Plant Species #1 (66.6666666666668 %) = 33.3333333333332 %

So, 33.3333333333332 % of total Bio-Cultural Diversity for Human Culture #1 remains.

ES = Ecosystem Services

@ = At

% = Percent

- = Minus

= -- Equals

AS #1=Animal Species #1

BCD HC #1=Bio-Cultural Diversity for Human Culture #1

Example # 2:

Let's say that animal species #1 is being used for clothing, food, shelter, and spirituality for human culture #1. Animal species #1 is being used for over 66.6666666666666668 % of human culture for human culture #1. Let's also say that animal species #1 has just gone extinct. This now means that due to animal species #1 going extinct, human culture #1 has just experienced a Bio-Cultural Diversity loss of over 66.6666666666666668 %, or a loss of cultural heritage by over 2/3 for the entire human culture being studied. We could also write this out as the following mathematical expression/equation:

ES @ 100 % - AS #1 (16.666666666666667 %) (4) = BCD HC #1 Remaining

ES @ 100 % - AS #1 (66.666666666666668 %) = 33.3333333333333332 %

This could also be written out as:

Ecosystem Services @ 100 % - Animal Species #1 (16.666666666666667 %) (4) = Bio-Cultural Diversity for Human Culture #1 Remaining

Ecosystem Services @ 100 % - Animal Species #1(66.666666666666668 %) = 33.3333333333333332 %

So, 33.3333333333333332 % of total Bio-Cultural Diversity for Human Culture #1 remains.

The above mathematical expressions and equations can be used to show the direct link and importance of both wildlife species diversity and the effects that declining wildlife species can have on human populations and cultures around the world. Essentially, what happens to wildlife also happens to humans. If wildlife species go extinct, so will some aspects of human culture and language. The above mathematical expressions/equations prove this direct correlation/link that humans are most certainly connected with nature, and heavily depend upon local and global ecosystem health. In Native American cultures, and many other Indigenous cultures around the world, our teachings tell us that we are one with the "Earth Mother." What we do to our "Earth Mother," we do to ourselves. We as Indigenous People have the understanding that "we are all related," "we are all connected," "we are all one," and "what happens to one of us will then happen to all of us."

All of the sacred plants and animals are our brothers and sisters. We depend upon them to be happy and healthy, so we can be happy and healthy. The Creator gave us the plants and animals to use for everyday life. Western science calls this "Ecosystem Services." This is an Indigenous teaching and philosophy that can now be scientifically proven (if using my mathematical equations and expressions listed above). I have now shown to you all how it is possible to use Indigenous teachings and philosophies, and how to integrate them into "Western sciences." In order for the world to see how special we as Indigenous Peoples are, I felt it was necessary to show you all from both perspectives. If humans are to continue to survive on this planet, you all will need to heed the teachings of the Indigenous Peoples of this planet, as their teachings (combined with Western scientific methodologies) could save us from global mass extinction. We are currently living in the "Anthropocene Epoch." Now is the time to wake up.

Bio-Cultural Environmental Use Theory

My theory states that the amount of environmental and ecological resources used will be dependent upon the needs of that particular society and culture. Factors and co-factors affecting environmental and ecological support/degradation within any given society/culture include:

- Cultural Practices/Societal Practices that either support or degrade wildlife, as well as the environment.

Example: African Ivory Trade. This trade is responsible for species becoming heavily endangered (such as the African Elephant-*Loxodonta*, and Black Rhinoceros-*Diceros bicornis*), due to peoples' beliefs that "horn ivory" has many medicinal benefits and healing properties to humans.

- "Socio-Economic Needs"-meaning natural resources being depleted or supported, based upon the socio-economic needs of a specific society or culture.

Example: In Brazil (South America's largest country), there is a high need for lumber. This is due to a high "country population." In turn, Brazilian timber companies are cutting down the Amazon rainforest at an unsustainable rate. This, in turn, destroys the Amazon's ecosystem, as well as depleting the world of its largest land-based oxygen supply. The Amazon is also home to many plants that "Western pharmaceutical companies" use to make medicine, not to mention that the Amazon houses the most diverse ecosystem in the world.

- Religious Practices and Spiritual Practices within a particular society/culture.

Example: The heavy use of cedar/sage in Native American culture (if not properly managed) could lead to cedar/sage plant population degradation. Populations of cedar/sage could also be supported through Native American culture, if a sustainable way of harvesting plants was identified.

Factors and co-factors could also affect ecological sustainability, support, and restoration

-Factors/co-factors that could affect ecological sustainability, support, and restoration are the same as above; i.e. cultural practices/societal practices, socio-economic needs, and religious practices.

This theory can be used as a "prediction of human-resource ecological modeling." If we know the populations of plants, animals, and the amount of resources that humans are using, then we can predict either positive/negative outcomes for the ecosystems, themselves. I recommend that "Ethnographic Field Methods" be utilized by a trained Ecological Anthropologist in order to determine whether the plants or animals being utilized are being supported or degraded, due to cultural and religious practices, and socio-economic needs. The "Bio-Cultural/Environmental Use Theory" is therefore a mixed methodological practice in order to evaluate, measure, estimate, and quantify both how resources are being utilized by humans (cultural use, socio-economic use, and religious/spiritual use), as well as establishing whether the total number of resources is being sustained or degraded by humans.

Public Broadcasting Service. (2014, October 24). Rhinoceros ~ Rhino Horn Use: Fact vs. Fiction-Nature. PBS. Retrieved September 17, 2022, from https://www.pbs.org/wnet/nature/rhinoceros-rhino-horn-use-fact-vs-fiction/1178/

Facts about Poaching. The Elephant Foundation. (n.d.). Retrieved September 17, 2022, from https://www.theelephantsociety.org/facts

Sandy, M. & Liste-Noor, S. *Why is the Amazon Rain Forest Disappearing?* Time. Retrieved September 17, 2022, from https://time.com/amazon-rainforest-disappearing/

Pember, M. A. (2020, November 27). *Native Americans Troubled by the Appropriation and Commoditization of Smudging.* Beauty Independent. Retrieved September 17, 2022, from https://www.beautyindependent.com/native-americans-troubled-appropriation-commoditization-smudging/

Model for Bio-Cultural Environmental Use Theory

1. If human consumption of ecological resources surpasses 40% of a resource's population (or total available resources), then human resource use is unsustainable and will degrade that ecosystem's population. This is because populations will need no more than 40% harvest in order to continue to grow in size.

2. If human consumption of ecological resources use is no more than 40%, then human resource use will be sustainable. This is because populations need a $^{+}60\%$ inclination rate to adequately grow in population size. Any more than 40% harvest or 40%>, the population will either neutralize (stay the same), or decline. This theory can be used for "predictive human-resource ecological modeling," and can also be used and integrated with "ecosystem service management," which is a way to express, quantify, and predict whether or not ecological resource use is either sustainable or unsustainable.

3. Why is this important? We are currently living in the "Anthropocene Epoch," where humans have caused almost all of the world's ecological issues/destruction. If we can identify the underlying causes of why humans are either supporting or degrading the ecosystems (through cultural, socio-economic, and religious/spiritual usages), then we can help to address, or even "reverse engineer," ecological destruction. At the very least, once the underlying causes are figured out, we can halt the destruction of Earth's resources. This is how my "Bio-Cultural/Environmental Use Theory" can be utilized in the world.

Human-Wildlife Resource Modeling (Predictive Modeling for Human-Wildlife Resource Use)

This form of modeling can be utilized in order to figure out how many overall resources would be utilized in the ecosystem, as a whole (Ecology). This form of modeling could also be utilized from a singular-species standpoint (either Botanical or Zoological) in order to figure out how many animal or plant resources would be individually utilized.

$$HRU = PH + AH$$

or

$$(OP - PR) + (OA - AR) = HRU$$

(For Overall Ecosystem Use)

Where:

$$OP - PR = PH$$
$$OA - AR = AH$$

This expression could also be written for individual-resource use (either plant- or animal-only).

$$HRU = PH$$
$$HRU = AH$$

or

$$HRU = PH (OP-PR)$$
$$HRU = AH (OA-AR)$$

and

OP = Original population of plants
PH = Plants being harvested
PR = Plants remaining
OA = Original population of animals
AH = Animals being harvested
AR = Animals remaining
HRU = Human resource use
+ = Adding or Positive
- = Minus or Negative
= -- Equals

Upon figuring out the amount of resources being utilized by humans, we could then look at the population numbers of the plants and animals, and then make a determination on a year-to-year basis whether or not these numbers would be sustainable for the population. We could also write this as an expression through time.

$$HRU = PH + AH \div 12 \text{ months in a year.}$$

HRU = Human resource use
PH = Plants being harvested
AH = Animals being harvested
= -- Equals
+ = Adding or Positive
÷ = Divided By
X = Multiplied By

The expression above would then give an "average" baseline for the amount of resources being used by humans per month, which would then give a finite number to be evaluated and studied on a month-to-month basis (short term), but could also be utilized in longer-term studies (year-to-year). We could even further the studies by including the overall human population to see exactly how many resources would be utilized on a month-to-month basis.

For example:

Say we have a city population of 100,000 people. Each individual harvests 10 plants and 10 animals per month for consumption. We could write this out as the following equation:

$$100{,}000 \text{ People} \times HRU = PH + AH \div 12 \text{ months in a year}$$

$$100{,}000 \text{ People} \times 20 \text{ (plants + animals being harvested)} \div 12 \text{ months in a year}$$

Answer: There would be 166,666.6667 resources used (both plants and animals) per month within the city's population.

This would be used for figuring out the combined amount of resources used by a human population for both plants and animals (Ecology-based approach).

We could also create a predictive model for human-wildlife resource use individually for either plant resources (Botanical) or animal resources (Zoological).

Example #1:

100,000 People X HRU=PH (Botanical) ÷ 12 months in a year

100,000 People X 10 (10 plants harvested) ÷ 12 months in a year

Answer: There would be 83,333.3333 resources used (Botanical) per month within the city's population.

Example #2:

100,000 People X HRU=AH (Zoological) ÷ 12 months in a year

100,000 People X 10 (10 animals harvested) ÷ 12 months in a year

Answer: There would be 83,333.3333 resources used (Zoological) per month within the city's population.

Now, let's add in more data to the above expression by integrating the number of permits for both plants and animals that are sold to citizens from the U.S. Forestry Service and the U.S. Wildlife Service in order to better determine how many plants or animals are being harvested (legally), and how this is affecting the overall population of either plants or animals on a year-to-year basis. Let's see the example below.

# PRHP = Number of Plant Resource Harvested for Permit
# ARHP = Number of Animal Resource Harvested for Permit
OP = Original population of plants
PH = Plants being harvested
PR = Plants remaining
OA = Original population of animals
AH = Animals being harvested
AR = Animals remaining
HRU = Human resource use
PIP = Permits issued for plants
TTSP = Total tree species population
PIA = Permits issued for animals
TASP = Total animal species population
X = Multiplied By
÷ = Divided By
= -- Equals
- = Minus or Negative
+ = Adding or Positive

Say we have a population of 100,000 people in the local area. The U.S. Forestry Service issues out 10,000 permits to harvest 1 Christmas tree per year (12 months in a year). Let's also say that there are 100,000 Christmas trees in the local area (total tree species population). This could be monitored through the following mathematical expression/equation:

Example #1:

100,000 People X PIP (# PRHP) ÷ OP = HRU

OP-HRU = PR

OP-PIP (#PRHP) = PR for Entire Year

PR = + or – Increase or Decrease in Population

100,000 People X 10,000 (Permits issued for plants) (1 Christmas tree legally harvested per year) ÷ 100,000 Christmas trees = 10,000 (10,000 Christmas trees harvested)

100,000 Christmas trees in the local area-10,000 Christmas trees harvested = 90,000 Christmas trees remaining in the local area

100,000 Christmas trees in the local area-10,000 (Permits issued for plants) (1 Christmas tree legally harvested per year) = 90,000 Christmas trees remaining for entire year

90,000 Christmas trees remaining = -10% decrease in population for entire year

The same process could be used for figuring out how to monitor animal populations. Say we have a population of 100,000 people in the local area. The U.S. Fish & Wildlife Service issues 10,000 permits to harvest 1 elk per year (12 months in a year). Let's also say that there are 100,000 elk in the local area (total animal species population). This could be monitored through the following mathematical expression/equation:

Example #2:

100,000 People X PIA (# ARHP) ÷ OA = HRU

OA-HRU = AR

OA-PIA (#ARHP) = AR for Entire Year

AR = + or – Increase or Decrease in Population

100,000 People X 10,000 (Permits issued for animals) (1 Elk legally harvested per year) ÷ 100,000 Elk = 10,000 (10,000 Elk harvested)

100,000 Elk in the local area-10,000 Elk harvested = 90,000 Elk remaining in the local area

100,000 Elk in the local area-10,000 (Permits issued for animals) (1 Elk legally harvested per year) = 90,000 Elk remaining for entire year

90,000 Elk remaining = -10% decrease in population for entire year

Special Thanks

All systems, GIS maps, modeling designs and uses, mathematical expressions/equations, theories, and biotechnologies/biotechnology templates were originally designed and created by Mr. Tyler Pounds. All mathematical expressions/equations were checked and verified for accuracy by Dr. Todd Walborn (PSY.D). I would also like to thank the State of Oklahoma and all of the Native American tribes that gave me access to their GIS and biodiversity data in the State of Oklahoma. These tribes include:

1. Wyandotte Tribe of Oklahoma

2. Ottawa Tribe of Indians of Oklahoma

3. Modoc Tribe of Oklahoma

4. Eastern Shawnee Tribe of Oklahoma

5. Cherokee Nation, Oklahoma

6. Peoria Tribe of Indians of Oklahoma

7. Choctaw Nation of Oklahoma

8. Seneca-Cayuga Tribe of Oklahoma

9. Miami Tribe of Oklahoma

10. Muscogee (Creek) Nation, Oklahoma

11. Quapaw Tribe of Indians, Oklahoma

12. Cheyenne-Arapaho Tribes of Oklahoma

13. Seminole Nation of Oklahoma

14. Potawatomi-Shawnee Nations, Oklahoma

15. Wichita-Caddo-Delaware Tribes, Oklahoma

16. Chilocco Indian School Lands of Oklahoma

17. Tonkawa Tribe of Indians of Oklahoma

18. Kaw Nation, Oklahoma

19. Kiowa-Comanche-Apache-FT. Sill Apache Tribes of Oklahoma

20. Ponca Tribe of Indians of Oklahoma

21. Chickasaw Nation, Oklahoma

22. Kickapoo Tribe of Oklahoma

23. Sac and Fox Nation, Oklahoma

24. Iowa Tribe of Oklahoma

25. Osage Tribe, Oklahoma

26. Pawnee Nation of Oklahoma

27. Otoe-Missouri Tribe of Indians, Oklahoma

I would also like to thank Kyle Trujillo of the Ancestral Lands Conservation Corps (Individual Placement Program), as well as the tribes of the American Southwest, Southwest Conservation Corps, and Save Our Bosque Task Force (SOBTF) for encouraging me to turn this research project into a book. These tribes include:

1. Navajo

2. Zuni

3. Cochiti-Pueblo

4. Isleta-Pueblo

5. Keresan-Pueblo

6. Hopi

7. Pima

8. Apache-White Mountain

9. Acoma-Pueblo

10. Apache-Chiricahua and Mescalero

11. Jemez-Pueblo

12. Keres-Pueblo (Western)

13. Laguna-Pueblo

14. Tewa-Pueblo

15. Havasupai

16. Mojave

17. Kamia-Kumeyaay

18. Paiute

19. Papago

20. Hualapai

21. Pima-Gila River

22. Navajo-Ramah

23. Yavapai

24. Apache-Western (Western Apaches)

25. Cocopah

26. Other Groups: Spanish American

Afterword and Acknowledgements

I am affiliated with the Eastern Shawnee Band of Oklahoma (Tecumseh's Band of Shawnee), and the Shawnee Tribe of Oklahoma (Cherokee-Shawnee). I was born in Dayton, Ohio (the Shawnee homeland), and I speak the Shawnee language fluently (as my grandparents taught it to me). Let me be clear that I do not currently qualify for tribal enrollment with my nation, due to tribal politics and technicalities that my tribe has set into place that currently disqualify me from tribal enrollment. My family was relocated from the Shawnee Ohio Reservation in Piqua, Ohio (1795) to the state of Oklahoma in 1830 by the order of President Andrew Jackson (Indian Removal Act), AKA The Trail of Tears. My family lived in Oklahoma and was enrolled on the original Shawnee rolls of Oklahoma during that time. There were many Shawnee families, including mine, who moved back to Ohio and Southern Indiana during the late 1800s, and around the year 1900. That was our people's homeland.

There was something called the "Indian Reorganization Act" that occurred during the 1940s which the government said that in order for the Shawnee tribe to retain federal recognition, all of the Shawnees in Oklahoma would have to register on a "state census," in order to retain tribal citizenship. To this day, in order for a person to be tribally enrolled, they have to be directly descended from one of the persons who had registered with the Shawnee 1940 Indian Reorganization Act Census. The problem is that my family moved back to Ohio and Indiana around 1890. Therefore, because none of my family lived in the State of Oklahoma during the time that the Shawnee Census was taken, I am permanently disqualified from ever receiving tribal enrollment (even though the Shawnee Tribe of Oklahoma and the Eastern Band of Shawnee of Oklahoma have my family listed on the original 1830 Shawnee rolls). I am descended from the McCoy, Wade, and Walborn families (on my biological father's side). They are all traditional Shawnee families. One must note that my former name is Tyler Walborn. My biological father is Todd Walborn.

My biological father and mother met and grew up together in the Greenville & Dayton Ohio area. When I was 3 or 4 years old my biological father (Todd Walborn) got a job at the University of Alabama (Tuscaloosa, Alabama). The reason of why my name is now Tyler Pounds is because my mother remarried (when I was five or six) in Tuscaloosa, Alabama where my stepfather (Jody Pounds) legally adopted me. Therefore, I had a legal name change on my birth certificate and in life. My biological father's parents were Louie and Barbara Walborn (Barbara Wade-McCoy). My grandmother Barbara's parents were Paul Wade and Marjorie McCoy. My mother was a Scots-Irish woman and my grandparents on my mother's side were Ralph and Joyce Magee (McGee). My mother's maiden name was Colleen Kay Magee (McGee). My stepfather was named Jody Garland Pounds. I also have one half-brother (from my mother and stepfather) named David Alexander Pounds. My mother, stepfather, half-brother all currently reside in the State of Alabama.

My grandparents and great grandparents were sent to Indian boarding schools in southeastern and central Indiana where they were slapped when they spoke their language, and my grandmother was raped as a young girl for being a "Shawnee Savage." It would be difficult for anyone to tell me I am not Shawnee when I was born in Ohio and I speak my tribal tongue fluently (it is dead in Oklahoma). I must also remind everyone that the "Certificate of Degree of Indian Blood" (CDIB) card system was specifically designed to destroy Native people from the inside out. None of the other Indigenous Peoples of the world have "Indigenous People cards." This only exists in the United States and Canada. Having a government ID does not make one Indigenous. It is how you live your life, how your speak your language, being connected to historical lands, going to ceremonies, and conducting traditional dances, such as I have since I was a young man.

Growing up around my Shawnee grandparents and great grandparents (on my biological father's side of the family), I was told before they died that the Indian boarding schools that they were all sent to were in the state of Indiana. The counties in which these "Indian boarding schools/mission schools" were located was in Jackson, Jefferson, and Randolph counties within the State of Indiana. My Shawnee grandparents and great grandparents were sent with Miami, Wyandot, and Delaware Indians to these schools. They also told me that they were born on "reed mats" in the "flatlands of Indiana," and that I was descended from "Tecumseh's Band of Shawnee," of which there were three chiefs: Tecumseh himself, who was the "High Chief," Grey Wolf, who was the "Medicine Man Chief," and Blue Jacket, who was the "War Chief." I was told by my grandparents and great grandparents that "Grey Wolf-Medicine Man Chief" was my great grandfather times six. I was also told that he signed the "Treaty of Greenville-1795." To this day, a pictograph of a Wolf's head with two peace pipes crossed can be seen. This was his signature. I was also told that Grey Wolf's wife was called "Penta," which means "flower" in the Shawnee language, and that she watched Tecumseh give his famous speech on the rock at the cave in Greenville, Ohio, as a little girl.

I was born in Dayton, Ohio and my grandparents and great grandparents (both on my father's and mother's sides) raised me in Greenville, Dayton, Piqua, Ohio. These were the original Shawnee tribal town epicenters. I was lucky to have grown up in traditional Shawnee homeland, as the Shawnee of Oklahoma will unfortunately never know the land of their ancestors, like I do. I am fortunate to have walked with our ancestors in our old tribal town areas. My grandparents and great grandparents told me that we are "Shuwaneesa," or the "Shawandassee," ayolas! We are Southern Sac and Fox's (Meskwaki). I was also told that the Shawnee of Ohio and Indiana were born from the "Great Serpent Mound," which is located in Ohio. This was where we held traditional stomp dance ceremonies for thousands of years before any settlers came into our lands demanding land and treaties.

Later on, my grandparents gave me a traditional Shawnee name, "Midwa Misqooney," which means "Fierce Heart" in the Shawnee language. I also fondly remember my family cooking "grape dumplings" and "blue corn grits" around a fire before we sang "Mingo songs." After we sang "Mingo songs," we would play the "Shawnee Ball Game," which included using a deer rawhide that had been filled with deer hair to kick across the field in order to score points. I also remember my grandparents growing a "corn garden" in their backyard for the "Shawnee Green Corn Ceremony," in which they grew four strains of corn in a giant circle corn crop patch to represent the sacred corn wheel that the Shawnee People use for ceremony. My grandparents "Corn Garden" consisted of Blue, Red, White, and Yellow Corn. They then harvested some of this corn from their "Corn Medicine Garden" to make a corn mash drink (sungasa) that was then used during the annual Shawnee "Green Corn Ceremony."

Since my grandparents' passing, I have been actively involved in the Shawnee Stomp Dance of Oklahoma, and the Keetoowah Cherokee Stomp Dance of Oklahoma (when I lived in Tahlequah, Oklahoma while attending Northeastern State University). I am also an active member of the Ghost Dance Society with the Shoshone, Paiute, and Bannock Tribes of Fort Hall, Idaho, as well as a supporting member of the Sun Dance Society of Fort Hall, Idaho (Shoshone-Bannock). One of my spiritual teachers, Clyde M. Hall, is a retired tribal judge and lawyer of the Shoshone-Bannock tribes of Fort Hall, Idaho. I should also mention that almost all of my TEK research has been with Indigenous People, but TEK is not exclusive to Indigenous People. Some TEK stems from farmers whose families have lived on the same farmsteads for 400+ years. This can be seen in New England, as well as other parts of the world. Technically, TEK knowledge can be held by any group of people who have had cultural and historical ties to a plot of land anywhere from a few hundred to a few thousand years. I should also mention that my research is heavily mathematical and science-based. I am combining both qualitative and quantitative methodologies in order to pursue these types of studies.

Last but not least, I want to give a warm shout out to the "Green Feather" family of the Shawnee Nation in Oklahoma. I went to Shawnee Stomp Dances while attending Northeastern State University. Don "Green Feather" took me to these Shawnee Stomp Dances. I hope you all are forever blessed and guided by the spirits of your ancestors. Aho!

About the Author/Illustrator/Front & Back Cover Artist:

This book was created, written, and illustrated by Tyler Pounds (formerly Tyler Walborn). Mr. Pounds is an Ecological Anthropologist/ Ethnobiologist. He has an M.A. in Anthropology (Ecological) from New Mexico State University with Graduate Minors in Fisheries, Wildlife, Conservation Ecology and Native American Studies. Mr. Pounds also holds a B.A. in American Indian/Native American Studies from Northeastern State University in Tahlequah, Oklahoma. Mr. Pounds is a National Science Foundation (NSF) summer scholar. Oddly enough, Mr. Pounds never graduated from high school. He holds a G.E.D. and wishes to be "a shining light" to people who come from disadvantaged backgrounds. He states, "it's never too late to fulfill your dreams in life: never giving up is of the utmost importance."

Mr. Pounds has worked alongside many federal, state-level, and non-profit agencies for wildlife conservation and cultural heritage. This includes the Bureau of Land Management, American Conservation Experience-Epic (Forestry Service), Americorps, Southwest Conservation Corps, Ancestral Lands Conservation Corps, Save Our Bosque Task Force, and the Naraya Cultural Preservation Council (NCPC). While working for these agencies, Tyler received first-hand experiences with wildlife management practices, and he is also a self-taught "Integrated Biologist." Tyler combines the fields of Anthropology and Wildlife Biology (Botany, Ecology, and Zoology) into an "integrated approach" fit for the 21st century. Tyler enjoys addressing issues that are directly related to the "Anthropocene Epoch," while also bridging the gaps between humans and wildlife. Mr. Pounds is currently employed by the "Institute of Natural Resources" at Oregon State University (PAC-12)-Corvallis, Oregon.

Anyone trying to reach Mr. Tyler Pounds can contact him below.

Email: tylerpounds1989@icloud.com

Follow me on these social media channels!

Facebook:
https://www.facebook.com/profile.php?id=100088237461903

Linkedin:
https://www.linkedin.com/in/tyler-pounds-87798125b/

Twitter:
https://twitter.com/tylerpounds1989

Instagram:
https://www.instagram.com/tpounds04/

Youtube:
https://www.youtube.com/channel/UCpQIY6TbJ0csL8r-lqH7PQQ

Goodreads.com:
https://www.goodreads.com/author/show/47839486.Tyler_Pounds

About the Editor:

This book has additional editing work that has been completed by Dr. Todd Walborn (Psy.D). Dr. Walborn completed his doctorate in Psychology at Wright State University in Dayton, Ohio. Dr. Walborn (Psy.D) can be contacted at the email address below.

toddwalborn@gmail.com

Front and Back Cover Artwork (Photographs)

Both the front cover and back cover artwork photographs were taken by Kimberly Roy. Kim holds a B.A. in American Indian/Native American Studies from Northeastern State University with minors in 2D and 3D Art (Tahlequah, Oklahoma). Kimberly Roy is an adventurous individual with a passion for photography and the great outdoors. Kim has also graduated from the Meredith Manor School of Horsemanship in Waverly, West Virginia.

Kimberly Roy can be contacted at the email address below.

kimberlyroysrainbowdreams@gmail.com

Sources Cited

Basehart, Harry W., 1974, Apache Indians XII. Mescalero Apache Subsistence Patterns and Socio-Political Organization, New York. Garland Publishing Inc., page 44

Basehart, Harry W., 1974, Apache Indians XII. Mescalero Apache Subsistence Patterns and Socio-Political Organization, New York. Garland Publishing Inc., page 50

Berkes, F. 1999. *Sacred Ecology: Traditional Ecological Knowledge and Resource Management*. Philadelphia, PA: Taylor and Francis Publishing Company.

Bridgewater, P. and Upadhaya, S. (2022, September 7). *What is Biocultural Diversity, and why does it Matter?* The Conversation. Retrieved September 24, 2022, from https://theconversation.com/what-is-biocultural-diversity-and-why-does-it-matter-168881

Buskirk, Winfred, 1986, The Western Apache: Living With the Land Before 1950, Norman. University of Oklahoma Press, page 178

Buskirk, Winfred, 1986, The Western Apache: Living With the Land Before 1950, Norman. University of Oklahoma Press, page 190

Camazine, Scott and Robert A. Bye, 1980, A Study of the Medical Ethnobotany of the Zuni Indians of New Mexico, Journal of Ethnopharmacology 2:365-388, page 378

Castetter, Edward F., 1935, Ethnobiological Studies in the American Southwest I. Uncultivated Native Plants Used as Sources of Food, University of New Mexico Bulletin 4(1):1-44, page 31

Castetter, Edward F., 1935, Ethnobiological Studies in the American Southwest I. Uncultivated Native Plants Used as Sources of Food, University of New Mexico Bulletin 4(1):1-44, page 43

Castetter, Edward F., 1935, Ethnobiological Studies in the American Southwest I. Uncultivated Native Plants Used as Sources of Food, University of New Mexico Bulletin 4(1):1-44, page 51

Castetter, Edward F. and Ruth M. Underhill, 1935, Ethnobiological Studies in the American Southwest II. The Ethnobiology of the Papago Indians, University of New Mexico Bulletin 4(3):1-84, page 65

Castetter, Edward F. and M. E. Opler, 1936, Ethnobiological Studies in the American Southwest III. The Ethnobiology of the Chiricahua and Mescalero Apache, University of New Mexico Bulletin 4(5):1-63, page 41

Castetter, Edward F. and M. E. Opler, 1936, Ethnobiological Studies in the American Southwest III. The Ethnobiology of the Chiricahua and Mescalero Apache, University of New Mexico Bulletin 4(5):1-63, page 44

Castetter, Edward F. and M. E. Opler, 1936, Ethnobiological Studies in the American Southwest III. The Ethnobiology of the Chiricahua and Mescalero Apache, University of New Mexico Bulletin 4(5):1-63, page 45

Castetter, Edward F. and M. E. Opler, 1936, Ethnobiological Studies in the American Southwest III. The Ethnobiology of the Chiricahua and Mescalero Apache, University of New Mexico Bulletin 4(5):1-63, page 53

Castetter, Edward F. and Willis H. Bell, 1951, Yuman Indian Agriculture, Albuquerque: University of New Mexico Press, page 122

Castetter, Edward F. and Willis H. Bell, 1951, Yuman Indian Agriculture, Albuquerque: University of New Mexico Press, page 201

Colton, Harold S., 1974, Hopi History and Ethnobotany, IN D. A. Horr (ed.) Hopi Indians. Garland: New York, page 319

Colton, Harold S., 1974, Hopi History and Ethnobotany, IN D. A. Horr (ed.) Hopi Indians. Garland: New York, page 324

Cook, Sarah Louise, 1930, The Ethnobotany of Jemez Indians, University of New Mexico, M.A. Thesis, page 22

Cook, Sarah Louise, 1930, The Ethnobotany of Jemez Indians, University of New Mexico, M.A. Thesis, page 28

Curtin, L. S. M., 1949, By the Prophet of the Earth, Santa Fe. San Vicente Foundation, page 78

Curtin, L. S. M., 1949, By the Prophet of the Earth, Santa Fe. San Vicente Foundation, page 88

Curtin, L. S. M., 1949, By the Prophet of the Earth, Santa Fe. San Vicente Foundation, page 96

Damian Carrington, *"The Anthropocene Epoch: Scientists Declare Dawn of Human-Influenced Age," The Guardian,* August 29, 2016, https://www.theguardian.com/environment/2016/aug/29/declare-anthropocene-epoch-experts-urge-geological-congress-human-impact-earth

Elmore, Francis H., 1944, Ethnobotany of the Navajo, Santa Fe, NM. School of American Research, page 26

Elmore, Francis H., 1944, Ethnobotany of the Navajo, Santa Fe, NM. School of American Research, page 38

Elmore, Francis H., 1944, Ethnobotany of the Navajo, Santa Fe, NM. School of American Research, page 62

Elmore, Francis H., 1944, Ethnobotany of the Navajo, Santa Fe, NM. School of American Research, page 68

Elmore, Francis H., 1944, Ethnobotany of the Navajo, Santa Fe, NM. School of American Research, page 75

Elmore, Francis H., 1944, Ethnobotany of the Navajo, Santa Fe, NM. School of American Research, page 83

Elmore, Francis H., 1944, Ethnobotany of the Navajo, Santa Fe, NM. School of American Research, page 96

Elmore, Francis H., 1944, Ethnobotany of the Navajo, Santa Fe, NM. School of American Research, page 97

Fewkes, J. Walter, 1896, A Contribution to Ethnobotany, American Anthropologist 9:14-21, page 17

Gifford, E. W., 1931, The Kamia of Imperial Valley, Washington, D.C., U.S. Government Printing Office, page 23

Gifford, E. W., 1931, The Kamia of Imperial Valley, Washington, D.C., U.S. Government Printing Office, page 24

Gifford, E. W., 1936, Northeastern and Western Yavapai, University of California Publications in American Archaeology and Ethnology 34:247-345, page 256

GIS Project: Tribal Lands of Oklahoma
Analyzing Species Richness and Biodiversity on Tribal Lands
Tyler Pounds-New Mexico State University-MA Anthropology (Ecological)
December 10, 2019
https://storymaps.arcgis.com/stories/380e04df620d48c793f312751aedd188

Hoagland, S. J. (2016). Integrating Traditional Ecological Knowledge with Western Science for Optimal Natural Resource Management. *IK: Other Ways of Knowing*, *3*(1), 1–15.
https://doi.org/https://doi.org/10.18113/P8ik359744

Hrdlicka, Ales, 1908, Physiological and Medical Observations Among the Indians of Southwestern United States and Northern Mexico, SI-BAE Bulletin #34:1-427, page 261

Hrdlicka, Ales, 1908, Physiological and Medical Observations Among the Indians of Southwestern United States and Northern Mexico, SI-BAE Bulletin #34:1-427, page 265

Jones, Volney H., 1931, The Ethnobotany of the Isleta Indians, University of New Mexico, M.A. Thesis, page 29

Jones, Volney H., 1931, The Ethnobotany of the Isleta Indians, University of New Mexico, M.A. Thesis, page 39

Jones, Volney H., 1931, The Ethnobotany of the Isleta Indians, University of New Mexico, M.A. Thesis, page 43

Jones, Volney H., 1931, The Ethnobotany of the Isleta Indians, University of New Mexico, M.A. Thesis, page 44

Kuh, G. D. (2008). High-Impact Educational Practices: What They Are, Who Has Access to Them, and Why They Matter, *Service Learning/Community-Based Learning,* Association of American Colleges and Universities. Retrieved October 21, 2022, from https://secure.aacu.org/AACU/PubExcerpts/HIGHIMP.html#:~:text=In%20these%20programs%2C%20field%2Dbased,solve%20problems%20in%20the%20community

Macbeth, S. (n.d.). *Participatory Action Research.* Participatory Action Research | Participatory Methods. Institute of Development Studies. Retrieved September 25, 2022, from https://www.participatorymethods.org/glossary/participatory-action-research

Maryellen Kennedy Duckett, National Geographic, "*Nature Needs us to Act-Now,*" March 4th, 2020.

Murphey, Edith Van Allen, 1990, Indian Uses of Native Plants, Glenwood, IL. Meyerbooks. Originally published in 1959, page 27

National Geographic, *Voices for Biodiversity, "Traditional Ecological Knowledge, Anthropology, and Climate Change,"* April 5, 2012, https://blog.nationalgeographic.org/2012/04/05/traditional-ecological-knowledge-anthropology-and-climate-change/

Pember, M. A. (2020, November 27). *Native Americans Troubled by the Appropriation and Commoditization of Smudging.* Beauty Independent. Retrieved September 17, 2022, from https://www.beautyindependent.com/native-americans-troubled-appropriation-commoditization-smudging/

Public Broadcasting Service. (2014, October 24). Rhinoceros ~ Rhino Horn Use: Fact vs. Fiction-Nature. PBS. Retrieved September 17, 2022, from https://www.pbs.org/wnet/nature/rhinoceros-rhino-horn-use-fact-vs-fiction/1178/

Rea, Amadeo M., 1991, Gila River Pima Dietary Reconstruction, Arid Lands Newsletter 31:3-10, page 5

Rea, Amadeo M., 1991, Gila River Pima Dietary Reconstruction, Arid Lands Newsletter 31:3-10, page 6

Rea, Amadeo M., 1991, Gila River Pima Dietary Reconstruction, Arid Lands Newsletter 31:3-10, page 7

Reagan, Albert B., 1929, Plants Used by the White Mountain Apache Indians of Arizona, Wisconsin Archeologist 8:143-61, page 159

Reagan, Albert B., 1929, Plants Used by the White Mountain Apache Indians of Arizona, Wisconsin Archeologist 8:143-61, page 160

Robbins, W.W., J.P. Harrington, and B. Freire-Marreco, 1916, Ethnobotany of the Tewa Indians, SI-BAE Bulletin #55, page 42

Robbins, W.W., J.P. Harrington, and B. Freire-Marreco, 1916, Ethnobotany of the Tewa Indians, SI-BAE Bulletin #55, page 64

Robbins, W.W., J.P. Harrington, and B. Freire-Marreco, 1916, Ethnobotany of the Tewa Indians, SI-BAE Bulletin #55, page 69

Russell, Frank, 1908, The Pima Indians, SI-BAE Annual Report #26:1-390, page 75

Russell, Frank, 1908, The Pima Indians, SI-BAE Annual Report #26:1-390, page 78

Russell, Frank, 1908, The Pima Indians, SI-BAE Annual Report #26:1-390, page 79

Russell, Frank, 1908, The Pima Indians, SI-BAE Annual Report #26:1-390, page 80

Sandy, M. and Liste-Noor, S. *Why is the Amazon Rain Forest Disappearing?* Time. Retrieved September 17, 2022, from https://time.com/amazon-rainforest-disappearing/

Seele, B. C., K. J. Esler, and A. B. Cunningham. 2019. Biocultural Diversity: A Mongolian Case Study. *Ecology and Society* 24(4):27. https://doi.org/10.5751/ES-11207-240427

Steggerda, Morris, 1941, Navajo Foods and their Preparation, Journal of the American Dietetic Association 17(3):217-25, page 222

Stevenson, Matilda Coxe, 1915, Ethnobotany of the Zuni Indians, SI-BAE Annual Report #30, page 60

Stevenson, Matilda Coxe, 1915, Ethnobotany of the Zuni Indians, SI-BAE Annual Report #30, page 70

Steward, Julian H., 1933, Ethnography of the Owens Valley Paiute, University of California Publications in American Archaeology and Ethnology 33(3):233-250, page 317

Stewart, Kenneth M., 1965, Mohave Indian Gathering of Wild Plants, Kiva 31(1):46-53, page 46

Swank, George R., 1932, The Ethnobotany of the Acoma and Laguna Indians, University of New Mexico, M.A. Thesis, page 26

Swank, George R., 1932, The Ethnobotany of the Acoma and Laguna Indians, University of New Mexico, M.A. Thesis, page 57

Swank, George R., 1932, The Ethnobotany of the Acoma and Laguna Indians, University of New Mexico, M.A. Thesis, page 62

Tangible and Intangible Cultural Heritage, Riches Resources, Published: November 27th, 2014. Accessed: August 9th, 2021 https://resources.riches-project.eu/glossary/tangible-and-intangible-cultural-heritage/

The Elephant Foundation. (n.d.). *Facts about Poaching*. Retrieved September 17, 2022, from https://www.theelephantsociety.org/facts

The Glossary of Education Reform. (2014, March 3). *Community-Based Learning*. Community-Based Learning Definition. Retrieved September 25, 2022, from https://www.edglossary.org/community-based-learning/

Train, Percy, James R. Henrichs, and W. Andrew Archer, 1941, Medicinal Uses of Plants by Indian Tribes of Nevada, Washington DC. U.S. Department of Agriculture, page 33, 34

Train, Percy, James R. Henrichs, and W. Andrew Archer, 1941, Medicinal Uses of Plants by Indian Tribes of Nevada, Washington DC. U.S. Department of Agriculture, page 123

UNESCO (2003) Convention for the Safeguarding of the Intangible Cultural Heritage. Paris: UNESCO.

UNESCO (n.d.) Guidelines for the Establishment of National "Living Human Treasures" Systems. Paris: UNESCO.

U.S. Department of the Interior. (2020, August 5). *Overview of TEK*. National Parks Service. Retrieved October 3, 2022, from https://www.nps.gov/subjects/tek/description.htm

U.S. Fish & Wildlife Service, "Traditional Ecological Knowledge for Application by Service Scientists," Ninepipe National Wildlife Refuge of Montana, October 2, 2018 https://www.fws.gov/nativeamerican/pdf/tek-fact-sheet.pdf

Vestal, Paul A., 1940, Notes on a Collection of Plants from the Hopi Indian Region of Arizona Made by J. G. Owens in 1891, Botanical Museum Leaflets (Harvard University) 8(8):153-168, page 159

Vestal, Paul A., 1952, The Ethnobotany of the Ramah Navaho, Papers of the Peabody Museum of American Archaeology and Ethnology 40(4):1-94, page 17

Vestal, Paul A., 1952, The Ethnobotany of the Ramah Navaho, Papers of the Peabody Museum of American Archaeology and Ethnology 40(4):1-94, page 39

Vestal, Paul A., 1952, The Ethnobotany of the Ramah Navaho, Papers of the Peabody Museum of American Archaeology and Ethnology 40(4):1-94, page 50

Vestal, Paul A., 1952, The Ethnobotany of the Ramah Navaho, Papers of the Peabody Museum of American Archaeology and Ethnology 40(4):1-94, page 52

Watahomigie, Lucille J., 1982, Hualapai Ethnobotany, Peach Springs, AZ. Hualapai Bilingual Program, Peach Springs School District #8, page 45

Weber, Steven A. and P. David Seaman, 1985, Havasupai Habitat: A. F. Whiting's Ethnography of a Traditional Indian Culture, Tucson. The University of Arizona Press, page 67

Weber, Steven A. and P. David Seaman, 1985, Havasupai Habitat: A. F. Whiting's Ethnography of a Traditional Indian Culture, Tucson. The University of Arizona Press, page 209

Weber, Steven A. and P. David Seaman, 1985, Havasupai Habitat: A. F. Whiting's Ethnography of a Traditional Indian Culture, Tucson. The University of Arizona Press, page 231

Weber, Steven A. and P. David Seaman, 1985, Havasupai Habitat: A. F. Whiting's Ethnography of a Traditional Indian Culture, Tucson. The University of Arizona Press, page 248

White, Leslie A, 1945, Notes on the Ethnobotany of the Keres, Papers of the Michigan Academy of Arts, Sciences, and Letters 30:557-568, page 562

Whiting, Alfred F., 1939, Ethnobotany of the Hopi, Museum of Northern Arizona Bulletin #15, page 32, 96

Whiting, Alfred F., 1939, Ethnobotany of the Hopi, Museum of Northern Arizona Bulletin #15, page 90

Whiting, Alfred F., 1939, Ethnobotany of the Hopi, Museum of Northern Arizona Bulletin #15, page 96